I0045889

Mines of
Cherry Creek, Ely Range and other Eastern Nevada Districts

by J. M. Hill

This is a photographic reproduction of the 1916 U.S. Geological Survey Bulletin 648 entitled "Notes on Some Mining Districts in Eastern Nevada."

Coverage includes Elko, Lander, White Pine, Lincoln and Clark Counties.

Nevada Publications
Box 15444
Las Vegas, Nev. 89114

Copyright ©1983 by Stanley W. Paher
ISBN 0-913814-65-2
Printed in the U.S.A.

CONTENTS.

3

ILLUSTRATIONS.

PLATE 1

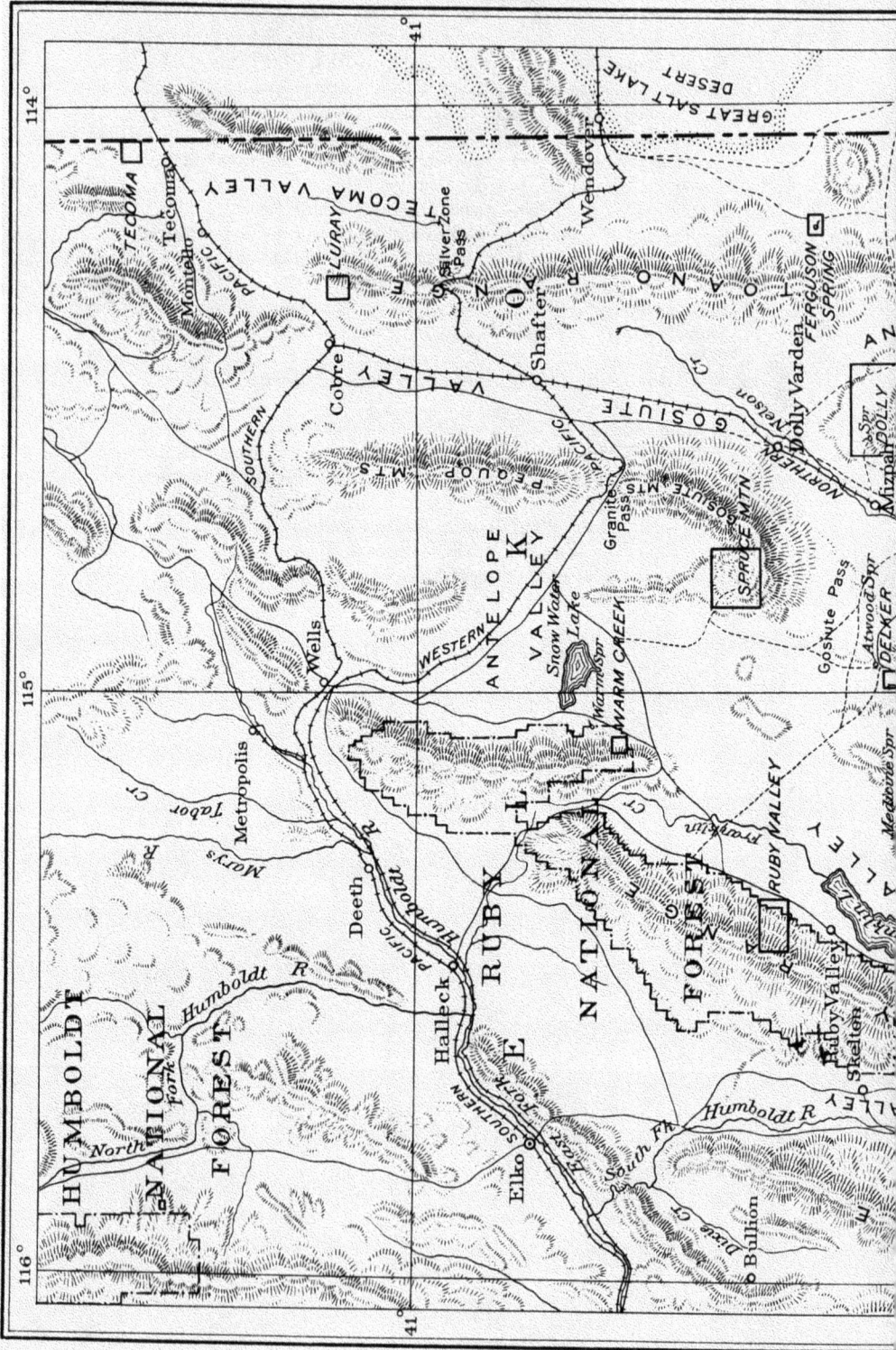

40

39

Deep Cr

NEVADA
UTAH

KERN MTS

K E R N

Monkey Spr

Antelope Hills

AURUM

Aurum

OSeola

FOREST

VALLEY

Tippett's

Tippett Spr

Antelope Spr

Schellbourne

N E

N A T I O N A L

Currant Spr

Gostute Y

Dry Bed

HUNT KINNEY

OPE RANGE

ANTELOPE RANGE

DUCK CREEK

D U C K   C R E E K

V A L L E Y

Taylor

Spr

Steptoe Cr

McGill

RANGE

SPRING

Cherry Creek

CHERRY CREEK

Egan Canyon

NEVADA

Duck

V A L L E Y

GRANITE

HUNTER

Steptoe

Summit Spr

Ely

WARD

NEVADA

Kimberly

Illipah

Hamilton

B U T T E   M T S

L O N G

B U T T E

W H I T E

BALD MTN

JOY

Ruby Pass

Warm Spr

RUBY

HUNT

Sherman

Desert

Alkali

DIAMOND

RANGE

Chokup Pass

WHITE PINE VAL.

Alkali Flat

White Pine Flat

FIRE PINE MTS

White R

Duckwater Cr

RANGE

SMOKY VALLEY

Eureka

EUR
NEV

ATLANTA
Atlanta
CEDAR RANGE
FORTIFICATION MTS
PATTERSON
LAKE VALLEY
Geyser
ELY RANGE
Bristol
Bristol Well
Pioche
OSPLA&SL
BRISTOL VALLEY
Sunnyside
PAHROC RANGE
PAHROC VALLEY
COAL VALLEY
GOLDEN GATE RANGE
Worthington PK
SIERRA
White River
NEVADA NATIONAL FOREST
WILLOW CREEK
IRON (IRWIN CANYON)
QUINN CANYON MTS
Quinn Canyon
ROAD VALLEY
Currant
Red Butte
LITTLE
LAKE

38° 114° 115° 116° 40 30 20 10 50 Miles

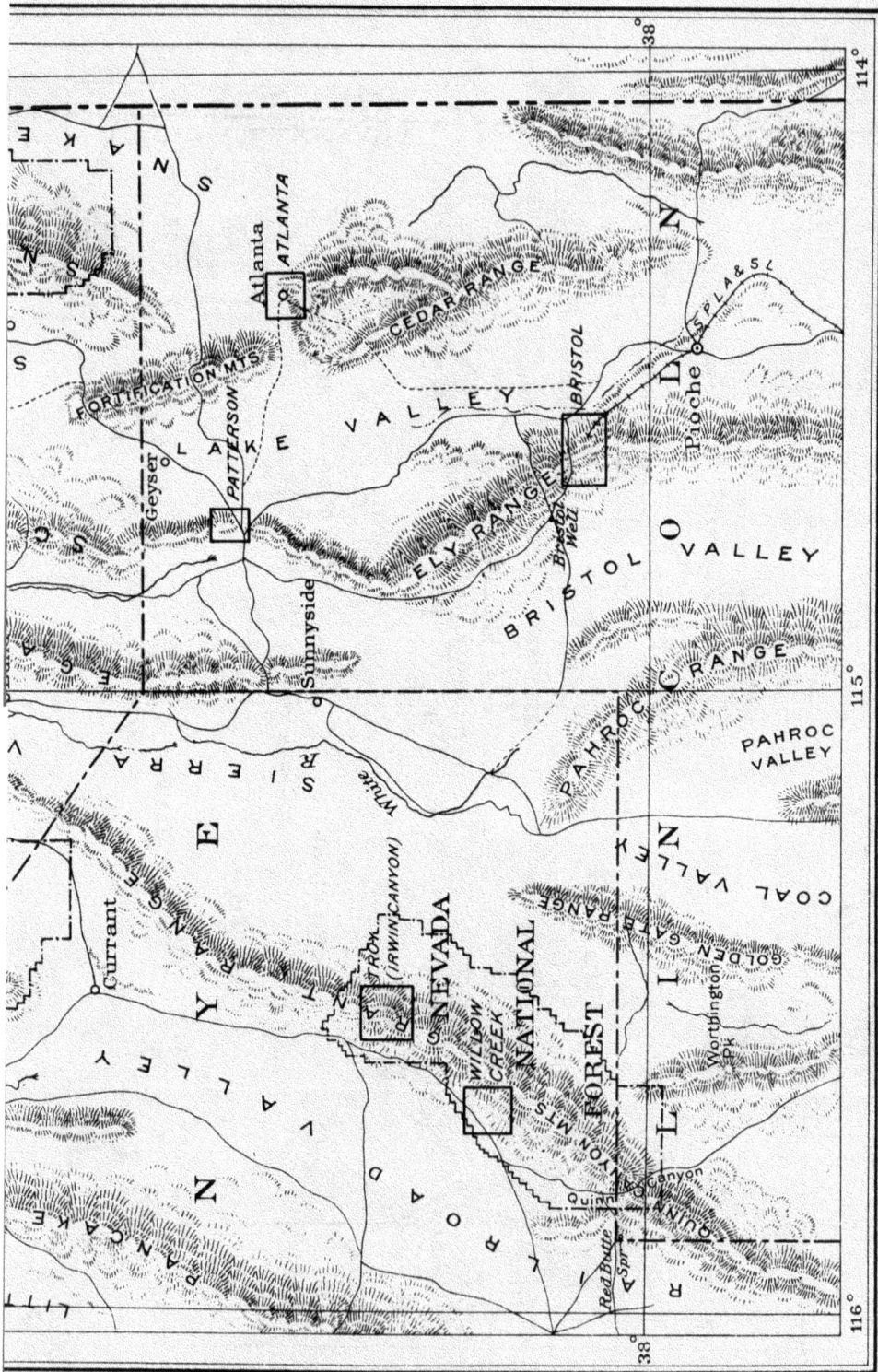

ENGRAVED AND PRINTED BY THE U.S.GEOLOGICAL SURVEY.

NOTES ON SOME MINING DISTRICTS IN EASTERN NEVADA.

By James M. Hill.

FIELD WORK AND ACKNOWLEDGMENTS.

During the summer of 1913 the writer continued a reconnaissance of scattered mining districts in Nevada, which he began in 1912.

The field work was begun in July, but was soon interrupted and was not resumed until the middle of September. Between that time and the first of November the writer visited 29 mining camps in Lander, Elko, White Pine, Nye, Lincoln, and Clark counties. As will be seen from an inspection of figure 1 the districts visited are widely scattered and much time was consumed in travel. On this account short stops only were made in the different camps, but it is believed that a general understanding of the major geologic features in each district was obtained. It is hoped that the conclusions here presented may be of some use as a guide to prospecting in this region.

It is a pleasure to record the numerous courtesies extended to the writer by the mining men of the region traversed. In every camp the information he desired was given to him and every facility was afforded to aid in his work. He wishes particularly to express his thanks to Mr. Harry Gentry, of St. Thomas; Mr. Creighton, of Elko; Messrs. Munter and Zeige, of Joy; Mr. Joseph Moore, of Currie; Messrs. Wales and Cannon, of Cherry Creek; Mr. Ornauer, of Hunter; Mr. David B. Gemmill, of Geyser; and Mr. Bent, of Bristol.

In the preparation of the material for this paper many of the more intricate petrographic problems have been solved by Mr. E. S. Larsen, and to Messrs. Knopf, Butler, and Umpleby the writer wishes to express his appreciation for their help and advice. He is indebted also to Mr. F. L. Ransome for a critical reading of the manuscript of this report and for kindly suggestions concerning many of the problems involved in the work.

ITINERARY.

The mines of the Tecoma district (No. 11, fig. 1) were visited in the middle of July. The writer was then detailed to other work, which was finished the middle of September. On September 15 the Ravenswood district (No. 15), Lander County, was reached, and the next two days were spent there. From Elko a 10-day trip was

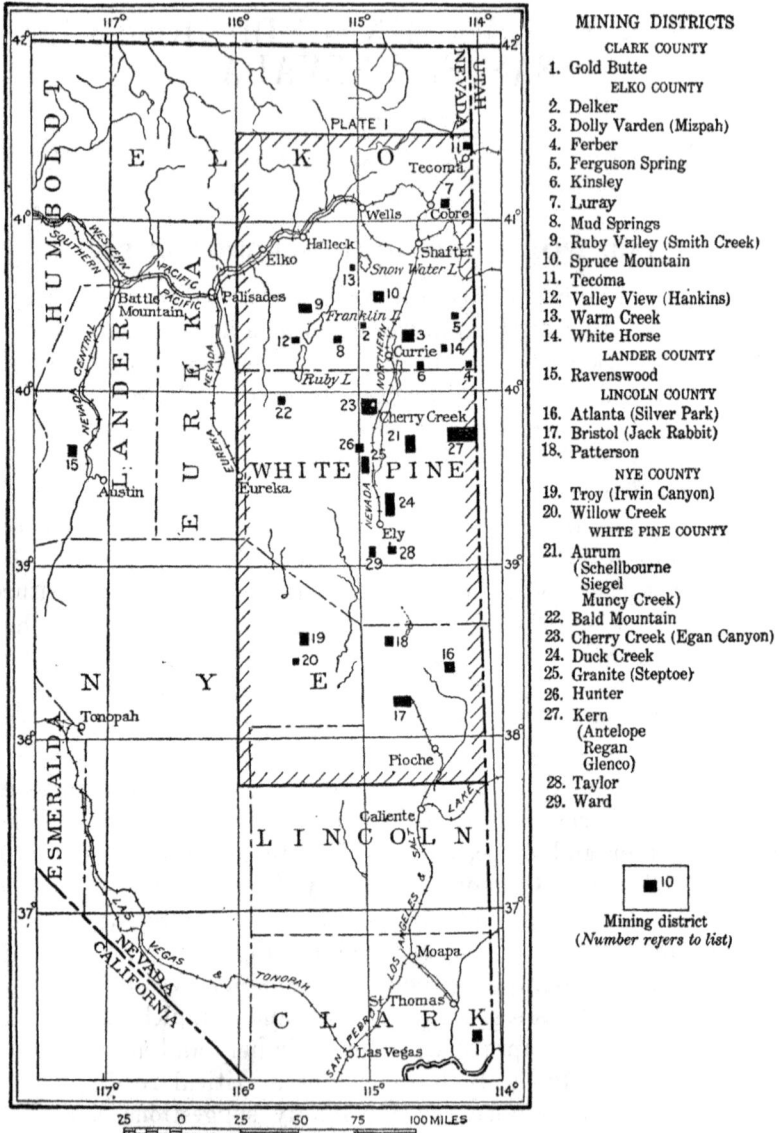

MINING DISTRICTS

CLARK COUNTY
1. Gold Butte
ELKO COUNTY
2. Delker
3. Dolly Varden (Mizpah)
4. Ferber
5. Ferguson Spring
6. Kinsley
7. Luray
8. Mud Springs
9. Ruby Valley (Smith Creek)
10. Spruce Mountain
11. Tecoma
12. Valley View (Hankins)
13. Warm Creek
14. White Horse
LANDER COUNTY
15. Ravenswood
LINCOLN COUNTY
16. Atlanta (Silver Park)
17. Bristol (Jack Rabbit)
18. Patterson
NYE COUNTY
19. Troy (Irwin Canyon)
20. Willow Creek
WHITE PINE COUNTY
21. Aurum
    (Schellbourne
    Siegel
    Muncy Creek)
22. Bald Mountain
23. Cherry Creek (Egan Canyon)
24. Duck Creek
25. Granite (Steptoe)
26. Hunter
27. Kern
    (Antelope
    Regan
    Glenco)
28. Taylor
29. Ward

■ 10

Mining district
(*Number refers to list*)

FIGURE 1.—Map of Nevada showing location of mining districts described in this report.

made in south-central Elko County, during which the mines in the Warm Creek district (No. 13), Ruby Valley (Smith Creek) district (No. 9), Valley View district (No. 12), Bald Mountain district (No. 22), Mud Springs district (No. 8), Delker district (No. 2), and

Spruce Mountain district (No. 10) were seen, Currie, on the Nevada Northern Railway, being reached on September 26. During the following 11 days, in a drive from Currie to Cherry Creek, by way of Antelope and Spring valleys, the mines of the Dolly Varden (Mizpah) district (No. 3), Ferguson Spring district (No. 5), White Horse district (No. 14), Kinsley district (No. 6), Kern Mountains (Eagle) district (No. 27), and the Aurum district (No. 21), which are scattered along the Schell Creek Range at Muncy Creek, Silver Creek, Siegel, and Schellbourne, were studied. Cherry Creek (No. 23), was reached on October 7, and the mines of the Granite and Hunter districts (Nos. 25 and 26) were next visited, Ely being used as a base.

From Ely a trip to the new gold strike at Willow Creek (No. 20) at the south end of Railroad Valley, in northeast Nye County, was made, and the prospects at Troy and Irwin Canyon (No. 19) were visited on the return trip. During the trip from Ely to Pioche the mines of the Ward (No. 29), Taylor (No. 28), Patterson Pass (No. 18), Atlanta (No. 16), and Bristol (No. 17) districts were studied. From St. Thomas, Nev., a trip to the Gold Butte camp (No. 1) was made late in October, and during the first of November the Grand Gulch mine, in northern Mohave County, Ariz., was studied. A description of this copper deposit by the writer has been published in Survey Bulletin 580.

In October, 1914, a short trip south from Wendover, Utah, was made to study the Ferber district (No. 4), in the southeast corner of Elko County, and to visit the tungsten mines on the southeast side of the Kern Mountains.

PREVIOUS DESCRIPTIONS.

The reports of the United States Geological Exploration of the Fortieth Parallel, particularly Volumes II and III, contain the best published geologic information relative to the southeastern part of Elko County. The topographic map which accompanied those reports has been freely used as a base for the present work, as it was found that no other maps were available for the purpose. Much of the geologic data gathered by that survey is used in this report, though some changes were found necessary in the interpretation of geologic evidence.

A great fund of information concerning the geology of the southeastern part of Nevada is contained in the annual reports and the final volumes of the United States Geographical and Geological Surveys West of the One Hundredth Meridian.

The history of Nevada from 1580 to 1888, contained in volume 25 of Bancroft's works, has supplied some details of the early mining

developments. From the biennial reports of the State mineralogist
of Nevada for the years 1866 to 1878, inclusive, many notes on early
development were obtained, and the semistatistical reports of J. Ross
Browne and R. W. Raymond have been searched for such information as bears upon the region under discussion. The annual sta-
tistical reports, "Mineral Resources of the United States," published
by the United States Geological Survey, contain many notes on mining development in recent years which are not found elsewhere.
The most recent general geologic report on the southern part of
Nevada issued by the Survey is entitled "The geology of Nevada
south of the fortieth parallel and adjacent portions of California,"
by J. E. Spurr, published in 1903 as Bulletin 208.

The following reports issued by the Survey deal more particularly
with the economic features of the several counties:

CLARK COUNTY.

Bain, H. F., A Nevada zinc deposit: Bull. 285, pp. 166, 169, 1906. Describes the
    Potosi mine in the Yellow Pine district.
Bancroft, Howland, Platinum in southeastern Nevada: Bull. 430, pp. 192–199,
    1910. Describes the platiniferous dikes of the Great Eastern and Key West
    region.
Hill, J. M., The Yellow Pine mining district, Clark County, Nev.: Bull. 540,
    pp. 223–274, 1914. Describes the zinc-lead, copper-gold, and gold deposits
    in the Spring Mountain Range tributary to Good Springs, Nev.
Ransome, F. L., Preliminary account of Goldfield, Bullfrog, and other mining
    districts in southern Nevada: Bull. 303, 1903. On page 79 mentions the
    gold-bearing veins of the Crescent district; on pages 63–76 describes the
    gold veins of Eldorado Canyon near Nelson and of the Searchlight district.

ELKO COUNTY.

Emmons, W. H., A reconnaissance of some mining camps in Elko, Lander, and
    Eureka counties, Nev.: Bull. 408, 1910. Describes the geology and ore
    deposits of the following districts in northern Elko County: Aura, Bullion,
    Centennial, Cornucopia, Good Hope, Tuscarora, Van Dusen.
Schrader, F. C., A reconnaissance of the Jarbidge, Contact, and Elk Mountain
    mining districts, Elko County, Nev.: Bull. 497, 1912. Describes the geol-
    ogy and ore deposits of three districts in the northern part of Elko County,
    near the Idaho line.

LANDER COUNTY.

Emmons, W. H., A reconnaissance of some mining camps in Elko, Lander, and
    Eureka counties, Nev.: Bull. 408, 1910. Describes the geology and ore
    deposits of the following districts: Bullion, Tenabo, Lander, Campbell,
    Cortez, Mill Canyon, Dean, Lewis, Mud Springs, and Hilltop.
Hill, J. M., Some mining districts in northeastern California and northwestern
    Nevada: Bull. 594, 1915. Describes the geology and ore deposits of the
    Battle Mountain, Copper Basin, Galena, Copper Canyon, Reese River,
    Skookum, New York Canyon, Washington, and Kingston districts.

NYE COUNTY.

Ball, S. H., A geologic reconnaissance in southwestern Nevada and eastern California: Bull. 308, 1907. Describes the occurrence of the ores of the following districts in central and western Nye County: Bare Mountain, Big Dune, Blakes Creek, Cactus Springs, Eden, Gold Bar, Gold Crater, Kawich, Gold Reed, Monte Cristo, Oak Springs, Reveille, Silverbow, Stonewall Mountain, Trappmans, and Wellington.

Becker, G. F., Geology of the quicksilver deposits of the Pacific slope: Mon. 13, 1888. Describes the occurrence of cinnabar at Belmont.

Emmons, W. H., and Garrey, G. H., Notes on the Manhattan district: Bull. 303, pp. 84–92, 1907.

Ransome, F. L., Preliminary account of Goldfield, Bullfrog, and other mining districts in southern Nevada: Bull. 303, 1907. Describes the ore deposits at Bullfrog and Rhyolite, on pages 40–62.

—— The geology and ore deposits of Goldfield, Nev.: Prof. Paper 66, 1909.

—— Round Mountain, Nev.: Bull. 380, pp. 44–47, 1910.

Ransome, F. L., Emmons, W. H., and Garrey, G. H., Geology and ore deposits of the Bullfrog district, Nev.: Bull. 407, 1910.

Spurr, J. E., Ore deposits of Tonopah and neighboring districts, Nev.: Bull. 213, pp. 81–87, 1903.

—— The ore deposits of Tonopah, Nev. (preliminary report) : Bull. 219, 1903.

—— Preliminary report on the ore deposits of Tonopah, Nev.: Bull. 225, pp. 89–110, 1904.

—— Developments at Tonopah, Nev., during 1904: Bull. 260, pp. 140–149, 1905.

—— Geology of the Tonopah mining district, Nev.: Prof. Paper 42, 1905.

WHITE PINE COUNTY.

Spencer, A. C., Geology and ore deposits of the Ely district, Nev.: Prof. Paper 96 (in press).

Weed, W. H., The copper mines of the United States in 1905: Bull. 285, pp. 116–117, 1906. Describes the copper prospects at Ely.

Weeks, F. B., An occurrence of tungsten ore in eastern Nevada: Twenty-first Ann. Rept., pt. 6, pp. 319–320, 1901.

—— Geology and mineral resources of the Osceola mining district, White Pine County, Nev.: Bull. 340, pp. 117–133, 1908.

—— Tungsten deposits in the Snake Range, White Pine County, eastern Nevada: Bull. 340, pp. 263–270, 1908.

THE REGION AS A WHOLE.

TOPOGRAPHY.

Striking features of the eastern part of Nevada are the arrangement of the mountain ridges along northward-trending axes and the wide, flat, barren valleys which lie between the ranges. This is well illustrated in the accompanying map of eastern Nevada (Pl. I).

RANGES.

The long, narrow northward-trending ranges are not everywhere of even height but are broken by narrow passes or low ridges into a number of parts, which have received different names. Thus the uplift shown on the left-hand side of Plate I is known as the Ruby Range in Elko County, as White Pine Range in White Pine County, and as the Grant and Quinn Canyon ranges in eastern Nye County. East of the White Pine uplift is the Egan Range, whose northern continuation in Elko County is the Spruce Mountain group and the Peoquop Range. East of the Egan Range the Schell Creek uplift of White Pine County is continued by the Ely and Highland ranges in Lincoln County. A spur which trends southeastward, known as the Fortification Range and Cedar Mountains, branches from the Schell Creek Range a few miles north of the Lincoln–White Pine County line. One of the most prominent and persistent uplifts in the region here discussed is that which extends along the east boundary of Nevada and is called the Toano Range in Elko County and the Kern Mountains and Snake Range in White Pine County.

The southern end of the Snake Range, culminating at Wheeler Peak, in southeast White Pine County, is the highest part of the ranges in this region, but the Schell Creek and Egan ranges, immediately west of Wheeler Peak, are also high and very rugged. The Ruby Range, in south-central Elko County, is very prominent, its peaks standing well above any of the neighboring mountains. Practically all the ranges shown on Plate I have steep eastern flanks, but the slopes to the west as a rule are more gentle. There are some places in all the ranges where the reverse conditions are seen, but in the main the above generalization holds true.

VALLEYS.

Lying between the mountains there are broad, flat-bottomed desert valleys which, like the ranges, have a northward trend. Eastern Nevada is somewhat better watered than the western and southern parts of the State, and near the mountains, particularly along the east base of the higher ranges, there are many cattle and hay ranches, which obtain their water supply from streams that have their source in the higher hills.

Huntington, White Pine, and Railroad valleys, which lie on the west side of the White Pine uplift, are separated by low, flat divides, and it is hard to realize that they are really not all the same basin. In this line of depressions the ranches are on both margins of the valleys, but as the Ruby and White Pine mountains are generally higher than the ranges west of the depression, most of the permanent streams rise in them, and it is only at a few localities that settlements or ranches are found on the west side of the valleys. There are no ranches on the east side of Ruby Valley, for most of the water in this vicinity rises in the Ruby Range. In Sierra and

Butte valleys, which are separated by a low divide, most of the ranches are on the east side, along the flanks of Egan Range. Steptoe Valley is not particularly well watered, though there are a few ranches near Currie, Cherry Creek, Steptoe, and Ely. Antelope Valley is barren. Spring Valley, which lies between Schell Creek and Snake ranges, is well watered, and there are many prosperous ranches along the base of both ranges, though the Schell Creek side seems to be better watered than the Snake Range side except at the south end, near Wheeler Peak.

GEOLOGY.

FEATURES OF GEOLOGIC STRUCTURE.

In eastern Nevada sedimentary rocks ranging in age from Cambrian to Permian have been faulted into their present positions along a series of northward-trending breaks of great magnitude. In general the major valleys seem to occupy the lines of faults. Beside the major series there are conspicuous evidences of eastward-trending faults, which were noticed in particular in the Schell Creek and Egan ranges. The sedimentary rocks have been at many places intruded by stocks and dikes of igneous rock, most of which closely approximate quartz monzonites or granite porphyries in their mineral composition. Two notable exceptions to the general similarity of the intrusive rocks were noted in this reconnaissance; one the gneissic biotite and muscovite granites of the Ruby Range and the other the muscovite-biotite granite of the Kern Mountains. Aside from these two exceptions the igneous rocks of one district differ but little in either looks or mineral composition from the rocks of any of the other districts. The stocks differ greatly in size; some occupying less than half a square mile, others covering many square miles. Large areas underlain by extrusive igneous rocks were found in remarkably few places in the part of eastern Nevada that was visited in 1913, but at a number of places flow rocks were noted.

Almost without exception, the ore deposits are localized about the centers of intrusion; indeed it is most surprising not to find either a stock or dikes of quartz monzonite or closely related rock in every camp.

SEDIMENTARY ROCKS.

Though fossils showing the age of the sedimentary rocks were not found in all the districts visited, it is believed that in the main the correlations given in this bulletin are fairly accurate. The correlation of many of the formations is based upon the similarity of sequence and of lithology of the rocks in the region covered by this reconnaissance with the rocks of areas in which the age of the formations has been definitely determined.

In eastern Nevada a large part of the sedimentary rocks are of early Paleozoic age, ranging from Cambrian to Devonian. In a few places early Carboniferous fossils were found, and along the east side of Ruby Valley some of the limestones are regarded as of Permian age.

CAMBRIAN ROCKS.

The Cambrian formations, which are very extensively developed in eastern and central Nevada, have been studied in considerable detail by Arnold Hague in the Eureka district, Eureka County.[1] The names applied by Hague, and later modified by Walcott,[2] appear to be well suited for all of eastern Nevada, as the formations seem to be very persistent over a wide area. Hague estimated that there are about 7,700 feet of quartzites, limestones, and shales of Cambrian age in the Eureka section, and divided the rocks into the following formations:

Prospect Mountain quartzite.—1,500 feet of brownish-white quartzites weathering dark brown, with intercalated thin layers of arenaceous and micaceous shales; beds lighter colored near top.

Prospect Mountain limestone.[2]—3,050 feet of light-bluish limestones, interbedded with light and dark limestones; usually crystalline and seamed with calcite veinlets; bedding indistinct; some lenses of shale.

Secret Canyon shale.—One thousand six hundred feet of yellow and brown argillaceous shale, passing into shaly limestone; weathers easily.

Hamburg limestone.—1,200 feet of dark-gray granular limestone; weathers rough and ragged. Is siliceous in places.

Hamburg shale.[2]—350 feet of yellow shale with abundant chert nodules in top; is variable in composition from arenaceous to calcareous.

Cambrian rocks that are regarded as representing the Lower Cambrian Prospect Mountain quartzite and the Middle Cambrian Eldorado limestone form most of the east side of the Egan Range from Currie southward to a point about opposite McGill. (See Pl. I.) The lower formation consists of the series of brownish weathering quartzite, with some interstratified micaceous shales, and the upper formation consists in part of white crystalline limestone and in part fine-grained gray limestone. The ore deposits of Egan Canyon, Cherry Creek, and Granite are inclosed in these rocks. In the Aurum and Patterson districts, on the east side of the Schell Creek Range, the same Lower and Middle Cambrian sequence is seen. In the Bristol

---

[1] Hague, Arnold, Geology of the Eureka district, Nev.: U. S. Geol. Survey Mon. 20, pp. 34–37, 1892.

[2] Because of the duplication of the names Prospect Mountain and Hamburg, Walcott in 1908 (Smithsonian Misc. Coll., vol. 53, p. 184) introduced the name Eldorado limestone to replace Hague's name "Prospect Mountain" limestone and Dunderberg shale to replace Hague's name "Hamburg" shale. Walcott's names have been adopted by the United States Geological Survey.

district what is thought to be the Cambrian Hamburg limestone contains the ore bodies. The dark limestones and argillites of the Kinsley district contain Cambrian fossils, and it is believed that the limestones and shales on the west side of the Kern Mountains are Cambrian. Cambrian fossils were found in the thin-bedded shales interstratified with limestones and quartzites in the Ravenswood district in Lander County. On the west side of the Grant and Quinn Canyon ranges in Nye County the dark-colored shales and limestones are believed to be in part Upper Cambrian and in part Lower Ordovician, as they are overlain by limestones and quartzites that are with little question referred to the Ordovician. The age of the quartzites and limestone of the Atlanta district is not definitely known, but they are thought to be either Cambrian or Ordovician.

On the eastern side of the Gold Butte district, in eastern Clark County, a series of quartzites and limestones overlying the Archean granite gneiss and granite is possibly the equivalent of the lower part of the Cambrian Tonto group of the Grand Canyon section described by Gilbert.[1]

ORDOVICIAN ROCKS.

The Ordovician section at Eureka comprises a thickness of 5,000 feet of sediments, divided into three formations, described by Hague[2] as follows:

Pogonip limestone.—2,700 feet of bluish-gray highly fossiliferous limestone that has a fine texture, weathers smooth, and has a fairly well marked massive bedding. Near the base it passes into the interstratified limestones, argillites, and arenaceous beds.

Eureka quartzite.—500 feet of white compact vitreous pure quartzite with indistinct bedding.

Lone Mountain limestone.—1,800 feet of limestone; dark gray, almost black; gritty at base, passing upward into bluish-gray limestone and finally into a light-gray siliceous limestone at top.

The white quartzite and dark limestones exposed on the east side of the Ruby Range at Smith Creek and Valley View are believed to be the equivalents of the Eureka and Lone Mountain formations, and the limestones and shales on the west side of the same range at Joy in the Bald Mountain district are probably of Pogonip age. In Nye County, at the Willow Creek and Troy (Irwin Canyon) districts of the Grant Range, limestones and quartzites regarded as of Ordovician age are represented in the higher mountains. Limestones that are believed to be the equivalent of the Pogonip are the prevailing rocks in the vicinity of the mines in the Duck Creek district and appear to be overlain by white quartzite that with little

[1] Gilbert, G. K., U. S. Geog. and Geol. Surveys W. 100th Mer. Rept., vol. 3, pp. 162–163, 184–185, 1875.
[2] Hague, Arnold, op. cit., pp. 47–62.

question represents the Ordovician Eureka quartzite. The limestones of the Taylor district may be Ordovician, though no very definite relations were noted to determine their age. A few small exposures of white quartzite along the large faults west of the mines at Bristol, Lincoln County, are thought to be the Ordovician Eureka quartzite.

Limestones that are older than Carboniferous are represented in the ridges east and south of Gold Butte, Clark County.

DEVONIAN ROCKS.

Devonian rocks are of considerable importance in eastern Nevada, consisting in the Eureka district of 6,000 feet of limestones to which Hague[1] gave the name Nevada limestone. The lower part of the Nevada limestone consists of light-gray crystalline, indistinct beds passing into brown, reddish-brown, and gray beds that are distinctly stratified and striped. The upper members are massive, bluish-black or dark-gray limestones. Intercalated bands of shale and quartzite are common near the middle of the formation.

Limestones determined by fossils as equivalent to a portion of the Nevada (Devonian) limestone of the Eureka district were seen in the Tecoma district, in Elko County, and the Hunter district on the west side of the Egan Range, in White Pine County.

CARBONIFEROUS ROCKS.

The oldest known Carboniferous formation in the Eureka district consists of 2,000 feet of black argillaceous shale and some red sandstones, to which Hague[2] gave the name White Pine shale. Although Hague assigned this formation to the Devonian period, some fossils found later show that it is early Carboniferous or Mississippian. Mississippian fossils were found in dark-blue and light-blue distinctly bedded limestones in a number of districts of eastern Nevada which are described in this report. Most of the sedimentary rocks of the Spruce Mountains and Dolly Varden Mountains are probably Mississippian, and it is believed that the limestones at Delker Butte are of the same age. The limestones on the east side of the Toano Range at Ferguson Spring, near the Nevada-Utah line, are Mississippian but are probably succeeded farther west in the main mountains by sediments as old as Cambrian. Mississippian fossils were collected from light-colored limestones a short distance north of Gold Butte, Clark County.

Pennsylvanian fossils were obtained from limestone in the Ward district, on the east side of the Egan Range, 16 miles south of Ely. Permian fossils were found in the limestones on the east side of

---

[1] Hague, Arnold, Op. cit., pp. 63–84.   [2] Idem, p. 68.

Ruby Valley, at Warm Creek and Mud Springs, and it is probable that the Pennsylvanian is represented at both of these places, though no fossils of that age were found.

TERTIARY DEPOSITS.

On the west side of the Ruby Range in Elko County (see Pls. I and III) a considerable area in Huntington Valley is underlain by partly consolidated Pliocene sandstones and conglomerates, which were referred to the Humboldt formation by the geologists of the Fortieth Parallel Survey. This formation grades from a rather coarse conglomerate near the base of the mountains to fine sand and silt near the axis of the valley. It shows a rough stratification and is only slightly consolidated. It is at some places difficult to distinguish between the Pliocene formation and the gravels and silts formed in the Quaternary, but in Huntington Valley the older gravels and silts have been slightly tilted and deeply eroded prior to the deposition of the more recent wash gravels.

QUATERNARY DEPOSITS.

The great desert valleys shown in Plate I (p. 20) are filled to an unknown depth with gravel and silt of Quaternary age. The centers of these valleys are underlain with fine sands, and in the lowest parts of most of them there are areas where fine clay silts are found. These playa lake beds are covered with water in the wet season but in the summer are dry and dusty. The silt supports no vegetation and, as a consequence, miniature dunes are formed in many of the basins. On the east side of Railroad Valley, between Currant Creek and Blue Eagle ranch, there are some fairly large sand dunes on the east side of the playa basin.

The material of the Quaternary fill becomes coarser toward the mountains, and there are gentle rises from the axes of the valley to the base of the ranges on both sides. Upon these long even gravel and sand slopes lie the outwash fans or cones which issue from the larger canyon mouths. The materials in the fans consist of coarse cobbles mixed with sand and gravel. The stream channels are intrenched in the gravels near the mouths of the larger canyons, but at the outer edge of the fans they are seldom more than shallow watercourses. The material in the alluvial fans is purely local, so that the character of the rocks in the mountains can be determined by a study of the cobbles and gravel found near the apexes of the fans.

Along Virgin River, in Clark County, the Quaternary is represented by an immense accumulation of partly consolidated, roughly stratified conglomerates and sandstones. Steep-walled canyons at least 200 feet deep have been cut in this formation, which resembles the deposits named Temple Bar conglomerate by Lee,[1] from their exposures at Temple Bar, near the mouth of Virgin River.

IGNEOUS ROCKS.

PRE-CAMBRIAN ROCKS.

A considerable area south of Gold Butte, in eastern Clark County, is underlain by a biotite granite gneiss of light-gray color and medium grain. In places it grades into biotite schist on the one hand or granite on the other. It is composed of quartz, orthoclase, microcline, some oligoclase, and widely varying amounts of green biotite. Gold Butte is composed of a coarsely porphyritic pinkish-gray granite, which is intrusive into the granite gneiss but does not appear to have penetrated the overlying sedimentary rocks. It is composed of orthoclase, microperthite, microcline, oligoclase, quartz, biotite, and hornblende. The accessory minerals are zircon and apatite and minor amounts of magnetite.

INTRUSIVE ROCKS.

Since the ore deposits of eastern Nevada seem to be so closely associated with intrusive granitic igneous rocks, it was surprising not to find some evidence of volcanism in each of the districts visited. No igneous rocks were seen in the vicinity of Warm Springs, on the east side of Ruby Range, but the prospects are not far from the immense stock of biotite granite in the main range. In the Mud Springs district there are apparently no igneous rocks, though quartz monzonite forms a large part of the Delker Buttes; nor were intrusives seen near the prospects at Ferguson Spring. So far as known, there are no intrusive rocks in the immediate vicinity of the mines of the Bristol district, Lincoln County.

Quartz monzonite.—Igneous rocks which closely approximate typical quartz monzonite in mineral composition are present in the following districts: Delker, Dolly Varden (Mizpah), Ferber, Kinsley, Valley View, and White Horse in Elko County; Ravenswood in Lander County; Troy (Irwin Canyon) and Willow Creek in Nye County; and Bald Mountain, Cherry Creek, and Ward in White Pine County. The quartz monzonite is typically granular, in some places somewhat porphyritic, and as a rule it occurs in roughly circular stocks of various sizes, though some dikes of this rock were noted. The quartz monzonite is usually siliceous, and in some stocks there is more quartz than feldspar in the rock. The feldspars include orthoclase, microcline, and plagioclase. The plagioclase varieties range from oligoclase-andesine to labradorite,

---

[1] Lee, W. T., Geologic reconnaissance of a part of western Arizona: U. S. Geol. Survey Bull. 352, pp. 17–18, 1908.

and as a rule are a little more abundant than the potassium feldspars. Dark-brown biotite and hornblende are the characteristic ferromagnesian minerals, but they vary in quantity and distribution, even in specimens from different parts of the same stock. Apatite and magnetite are the common accessory minerals and zircon is present in some places.

In general there has been some contact metamorphism of the limestones intruded by quartz monzonite, but evidence of extensive changes of this character is lacking in most of the districts visited, and in some places the sedimentary rocks are practically unaltered. In fact the lack of extensive metamorphism is a striking feature of the contacts seen in eastern Nevada.

The ore bodies of the districts where quartz monzonite is present were apparently formed after the consolidation of the magma; but it is believed that the interval between intrusion and metallization was comparatively short. Where veins occur in the quartz monzonite the solutions that deposited the ores have also altered the igneous rock. In many of the districts visited the alteration has not been extensive, yet the character of the change is the same in all places. The feldspars and ferromagnesian minerals are altered; the former to masses of sericite and calcite with some quartz, and the latter to chlorite or quartz and iron oxide.

Granite porphyry.—Some of the intrusive stocks do not carry sufficient plagioclase feldspar to be classed as quartz monzonites, yet these rocks are thought to be simply differentiates of the same quartz monzonite magma from which the quartz monzonites were derived and of the same age. Specimens of intrusive rock collected in a number of districts, but too much altered for definite determination of the feldspars, have been called granite porphyries rather than quartz monzonite porphyries. The intrusive rocks in the Spruce Mountain, Tecoma, Patterson, Aurum, Duck Creek, Hunter, Steptoe (Granite), and Taylor districts were called granite porphyry.

In most of the districts the granitic rocks have a porphyritic texture, but in the large stocks of the Granite and Aurum districts part of the rock is granular. Many of the bodies of granite porphyry are dikes rather than stocks—a fact that may in part account for the rather persistent porphyritic texture of so much of this rock. Quartz and orthoclase feldspars are the principal constituents of these rocks, and they occur both as the phenocrystic minerals and in the groundmass. Oligoclase is characteristic where the lime-soda feldspars can be determined, but in many of the thin sections of this type of rock the feldspars are too much altered for determination. Biotite is the most common ferromagnesian mineral, and hornblende is present in some of the granite porphyry masses. Apatite and magnetite are the accessory minerals most commonly found.

The granite porphyries, like the quartz monzonite porphyries, have been of importance in the formation of the ores. They have caused contact metamorphism of the sedimentary rocks, and deposits of similar ore minerals are found near bodies of the two types of rock. The alteration of the granite porphyry by mineralizing solutions is in all respects similar to that seen in the bodies of quartz monzonite.

Muscovite granite.—The large intrusive stock of the Kern Mountains is a fine-grained light-colored muscovite granite, which in some places contains enough biotite to be called a muscovite-biotite granite. It is composed essentially of quartz, orthoclase, and muscovite but in places carries a considerable amount of greenish-brown biotite. Magnetite is the most abundant accessory mineral. In the main mass the constituent minerals are about one-sixteenth inch in average size and show no crystal faces. Near the margin of the stock the mica flakes are locally well developed and at the contact the rock has been crushed to form a granitic schist. Some large bodies of muscovite pegmatite are found in the granite and in the sedimentary rocks near the contact. At one place a pegmatitic dike was seen that appeared to grade into a pure quartz vein. This rock is clearly intrusive into sedimentary rocks of supposed Cambrian age.

Biotite granite.—Most of the northern part of the Ruby Range, bordering on Ruby Valley, is composed of a light-colored fine-grained granite, which consists essentially of quartz, orthoclase, and brownish-green biotite, though in places microcline, oligoclase, muscovite, and light-green hornblende are also present. Apatite is the most abundant accessory mineral, but some magnetite is seen in all the thin sections examined. Much of the granite in this stock shows a distinct gneissic banding, yet considerable bodies of the rock have a roughly porphyritic texture and some pegmatitic granite is also present. This rock was considered of Archean age by the geologists of the Fortieth Parallel Survey, but it is clearly intrusive into sedimentary rocks that are considered Ordovician.

Other intrusive rocks.—In most of the stocks of quartz monzonite there are small dikes of fine-grained light-colored aplite, and in not a few places basic dikes cut the intrusives. These end-product dikes are so generally recognized that they need no discussion. In the Spruce Mountain district there are some dikes and a fairly large stocklike body of diorite, which are thought to be differentiates of the magma from which the granite porphyry was formed. Some facies of the quartz monzonite of the Dolly Varden Mountains are dioritic, containing much more hornblende and plagioclase than the normal rock. At Ferber a gabbroic rock consisting essentially of augite and andesine with some orthoclase and quartz is possibly a differentiate of the quartz monzonite magma.

A single specimen of diabase was obtained in the Cherry Creek district. In the field this rock was thought to be part of a series of metamorphosed shales and quartzites, and its true character was only determined when a thin section was examined microscopically. Its relations to the other rocks are not known. It is a dense black rock consisting of plagioclase feldspar, brown augite, biotite, and a little magnetite.

At the Melbourn mine, in the Willow Creek district, Nye County, there are some narrow dikes of andesite porphyry that are called diorite by the prospectors.

Age of the intrusions.—With the possible exception of the biotite granite of the north end of the Ruby Range, the muscovite granite of the Kern Mountains, and the diabase of the Cherry Creek district, all the intrusive rocks seen in the districts described in this report are thought to be of about the same age. They intrude sedimentary rocks ranging in age from Cambrian to Mississippian, but so far as known they are not in contact with younger rocks in this region. There is therefore no basis for a definite determination of their age. It is believed, however, from the fact that they have nearly the same mineral composition as many of the intrusions in Nevada, Utah, and California, which have been assigned to the late Mesozoic or early Tertiary, that they should be correlated with one or the other of these periods of intrusion. It is established that the volcanism of the Sierra Nevada took place near the close of the Mesozoic,[1] and it is nearly as well established that much of the intrusion of the intermediate rock types in Utah took place in the Tertiary. It may be that in Nevada intrusions of the type of rocks known as quartz monzonite and granite porphyry occurred during both periods of volcanism.

The age of the two stocks of potassium-rich granite, the biotite granite of the Ruby Range, and the muscovite granite of the Kern Mountains is more uncertain. These rocks have an entirely different appearance from the rocks of intermediate composition, whose correlation is somewhat more assured. At some places the granites have a decided porphyritic texture, but in general they are granular and at many places they show a gneissic banding, which might indicate considerable age. In the Ruby Range the biotite granite is intrusive into and has metamorphosed rocks that are believed to be Ordovician, and in the Kern Mountains the muscovite granite intrudes sediments regarded as Cambrian and possibly also Mississippian limestones. In the Ruby Range there is a slight suggestion

---

[1] Lindgren, Waldemar. Metallogenetic epochs: Canadian Min. Inst. Jour., vol. 12, 1909.

that the biotite granite of the northern part of the mountains may grade into the quartz monzonite mass at Harrison Pass, but the exact relations were not determined in the reconnaissance upon which this report is based. The writer ventures the suggestion, however, that these granites are differentiates of the magma from which the quartz monzonite and granite porphyry masses were formed.

EXTRUSIVE ROCKS.

Distribution.—Extrusive igneous rocks are not widely distributed in eastern Nevada, and as a rule underlie comparatively small areas in this part of the State. In southeastern Elko County, however, a considerable area northeast of the Schell Creek Range, extending to the Dolly Varden and Kern mountains, known in part as the Antelope Hills (see Pl. I), is for the most part underlain by flow rocks. Andesites of dull brown, yellow, and gray are most widely distributed in this area, being exposed on the west and south sides of the Dolly Varden Mountains, west of Kinsley Canyon, and north of Siegel Canyon, in the Aurum district. In general the rocks are glassy porphyries, but some flows have been altered to earthy rocks. At Dolly Varden a few exposures of rhyolite underlie the andesites. At Ferguson Spring and at Ferber, on the east side of the Toano Range, south of Don Don Pass, in Elko County, there are remnants of red and gray rhyolite flows. The exposures as a rule cover only a small area and are thin.

In the Ravenswood district, Lander County, rhyolitic breccia and flow rocks outcrop along the east base of the Shoshone Range, overlying gravels of the Tertiary Truckee formation. On the summit of the range there are some small remnants of dark-brown glassy andesite flows.

The Fortification Range and the Cedar Mountains, in Lincoln County, are composed largely of a complicated sequence of rhyolite and andesite tuffs and flows that Spurr has described in some detail. (See pp. 114–116.) In the vicinity of the Atlanta mines white biotite rhyolite tuffs are overlain by thin flows of red rhyolite.

South of Willow Creek district, in Nye County, Spurr reports a considerable area underlain by rhyolite. The lower part of the series, exposed at the summit of the south wall of Willow Canyon, is a gray porphyritic rhyolite with very little glass. In Troy Canyon, about 20 miles north of Willow Canyon, there are small exposures of a much-altered red glassy rhyolite.

Southwest of Gold Butte, in Clark County, the pre-Cambrian granite gneiss is overlain by what appear to be basalts, and somewhat similar though much fresher basalts are intercalated with the Quaternary gravels on the west side of Grand Wash, east of this district.

Age.—With the exception of the basalts of the Gold Butte district that are interbedded with Quaternary gravels, all the flow rocks of eastern Nevada are thought to be of Tertiary age, corresponding to the very extensive volcanism which has covered so much of western Nevada, southern Idaho, northeastern California, southwestern Oregon, and southwestern Utah with deep accumulations of extrusive rocks.

ORE DEPOSITS.

LOCATION AND CHARACTER.

With the exception of the ore body at the Atlanta mine, in Lincoln County, and the veins in the pre-Cambrian granite of Gold Butte, all the ore deposits seen by the writer during the summer of 1913 occur in sedimentary rocks, ranging in age from Cambrian to Permian, or in granular igneous rocks which have intruded the sedimentary beds. Most of the ore deposits occur in the sedimentary rocks; indeed, it is exceptional to find strongly mineralized veins in the igneous rocks in the districts described in this report. The mineralization of the ore bodies, except those at the Atlanta mine and at Gold Butte, as noted, is believed to have taken place soon after the time of intrusion. It appears that the mineralization about the centers of intrusion of the quartz monzonite and the related granite porphyry magmas was more intense than that about the centers of intrusion of the muscovite and biotite granite. It is believed that the ore deposits of all the districts except those just named were formed in either the "late Mesozoic" or the "early Tertiary" period of mineralization distinguished by Lindgren.[1]

At the Atlanta mine, in Lincoln County, the free-milling gold ore is found in a fault breccia involving a biotite rhyolite tuff of Tertiary age and is more closely related to the "late Tertiary" period of mineralization than to the earlier periods, to which all the rest of the ore bodies have been referred.

The age of veins in the pre-Cambrian granite gneiss and granite of the Gold Butte district is uncertain, though from lack of evidence to the contrary they are probably best assigned to the pre-Cambrian period of mineralization.

The more important ore deposits in the districts of eastern Nevada visited by the writer in 1913 are inclosed in sedimentary rocks and can be divided into three large types—replacement deposits, contact-metamorphic deposits, and veins. Not all of these types are represented in each district, and in some places only one type is seen. Second in importance are the deposits inclosed entirely in granular

---

[1] Lindgren, Waldemar, Metallogenetic epochs: Canadian Min. Inst. Jour., vol. 12, pp. 102–113, 1910 ; Econ. Geology, vol. 4, pp. 409–420, 1909.

igneous rocks, of which quartz veins are the most widely distributed type, though two deposits of disseminated character were seen. The Atlanta ore body, in part inclosed in Tertiary volcanic rocks, was the only deposit of this type seen in the course of the reconnaissance of the twenty-seven mining districts described in this report.

DEPOSITS IN SEDIMENTARY ROCKS.

REPLACEMENT DEPOSITS.

There are replacement deposits in eighteen of the twenty-nine mining districts visited in 1913. Most of the deposits are oxidized lead ores carrying varying amounts of silver and some carry also zinc and copper. All the deposits of this type are closely related to fissures, and many of them are fissure replacements, although not a few bedded replacements were seen. In general, the replacement ore bodies, particularly those carrying more lead than copper, are at some distance from the intrusive rocks, but in the Hunter and Ward districts the igneous rocks as well as the sedimentary rocks have been replaced by lead minerals.

As a rule, the exploitation of these deposits has not penetrated the sulphide zone, and cerusite, anglesite, and residual galena are the principal metallic minerals encountered in the mining. Smithsonite occurs in varying amounts in some deposits but is absent from many of them. Copper carbonates and copper-bearing iron oxide, though not universally present, are very commonly seen in the oxidized ores and in some districts form rather large ore bodies. In the Aurum district there are some peculiar silver ores in the oxidized zone, which consists almost entirely of pyrolusite carrying a little lead carbonate and silver, which is believed to be in the form of chloride.

In the Ward district, where the original sulphides have been cut in the mining, galena, sphalerite, pyrite, and chalcopyrite are the metallic minerals. At the Siegel mine, in the Aurum district, arsenical pyrite and galena are the original sulphides, and at the Signal mine pyrrhotite and pyrite are found in the unoxidized ores.

Replacement deposits, valuable chiefly for lead and silver, have been mined in the Dolly Varden, Mud Springs, Spruce Mountain, Ferber, Tecoma, Bristol, Patterson, Aurum, Duck Creek, Steptoe (Granite), Hunter, and Kern districts. In the Ward and Taylor districts, and in some of the deposits in the Aurum and Atlanta districts, the silver in the replacement ores is more valuable than the lead. Copper, which is accompanied by some silver and lead, is the most valuable metal of the replacement deposits of the Gold Butte, Ferguson Spring, and Bald Mountain districts and in some of the deposits of the Bristol district. In the replacement ores of the Willow Creek district the silver is more valuable than the copper. The slightly developed prospects of the Warm Creek district give promise of being more valuable for zinc than for lead.

In 9 of the 29 districts described in this report there are ore deposits containing minerals characteristic of contact metamorphism, but in two of these districts they are of little importance. In view of the number and the size of the intrusive bodies in eastern Nevada the lack of extensive contact metamorphism is peculiar. Much of the limestone near the intrusive bodies has been recrystallized, and in the recrystallization has as a rule been bleached of any decided color it may originally have had. The bodies of lime silicate rock, which are typical of the borders of so many intrusive masses, are, however, developed to only a slight extent in the region covered during 1913. Large bodies of the dark lime silicates are rarely met, and the few small lenslike bodies of contact-metamorphic minerals that are found near the intrusives are relatively inconspicuous. Among the lime silicate minerals noted during this reconnaissance light-green garnet is the most common and is followed in order of abundance by tremolite, diopside, dark-brown garnet, epidote, biotite, phlogopite, and clinozoisite. Quartz and calcite are present in varying amounts in these masses. Most of the contact-metamorphic deposits are valuable chiefly for their copper, but some contain principally lead and zinc, and one deposit of this type carries tungsten and bismuth.

Deposits of contact-metamorphic origin that are principally valuable for copper, but that carry some gold and silver, are found in the Delker, Dolly Varden, Ferber, Kinsley, Ruby Valley, Spruce Mountain, Aurum, and Bald Mountain districts. Most of these deposits have been but slightly developed, and in no district was absolutely unoxidized sulphide ore seen. The ores of the copper-bearing contact-metamorphic deposits are copper pitch ore, limonite, malachite, chrysocolla, and azurite. Residual kernels of pyrite and chalcopyrite are found here and there, and in a few mines a little galena and sphalerite, with their oxidation products, are found with the copper ores. At Smith Creek, in the Ruby Valley district, galena and sphalerite, with cerusite and smithsonite, are more abundant than the copper minerals in some of the lenses of lime silicate rock, though in other bodies the copper minerals predominate. A few small bodies of lime silicate rock in the Ward district contain, besides garnet and quartz, some galena and pyrite.

A peculiar occurrence of scheelite and bismuth was seen at the Valley View prospects, on the east side of the Ruby Range. These ores occur in limestone intruded by narrow, much-altered dikes, along which the limestones have been changed to small lenses consisting of clinozoisite, phlogopite, calcite, scheelite, bismuthinite, and native bismuth.

VEINS.

Veins that cut the sedimentary rocks were found in the Ravenswood district, in Lander County, in the Willow Creek district, in Nye County, and in the Cherry Creek and Granite districts, in White Pine County. In all these districts the veins occur in rocks believed to be of Cambrian age and most of them cut across the bedding, although at some places the veins and beds coincide in strike and dip. In the Steptoe (Granite) and Willow Creek districts the veins have not been extensively developed and free gold is the only valuable constituent of the ore, but here and there specks of pyrite are visible in the less oxidized ore. The gangue minerals are quartz and calcite, but in the Willow Creek district the rich "picture ore" is always associated with a green talc that seems to have been formed after the deposition of the quartz. In the Troy district large lenses of white quartz lie parallel to the bedding of the inclosing shaly limestones. The ores originally mined contained silver chloride associated with blue and green copper-bearing minerals and probably some of the rich secondary silver minerals. At a shallow depth the miners cut the water level, below which quartz, much of which is barren, contains some pockets of dark sphalerite and pyrite that chemical tests show to contain both copper and silver. Some of the veins in the Ravenswood district cut the Cambrian sediments; others lie parallel to the bedding and are generally of lenslike form. White and smoky quartz and barite are the gangue minerals, and are associated with chalcopyrite, antimonial galena, and tetrahedrite, which are said to carry some gold and silver. The sulphides are seen at the surface, but oxidation occurs to some extent to a depth of at least 240 feet. The oxidation products are chrysocolla, minor amounts of copper carbonates, anglesite, cerusite, and pyromorphite.

In the Cherry Creek district the veins have been more extensively exploited than elsewhere. They show a gradation from barren-looking quartz carrying a little pyrite, galena, and free gold to veins composed largely of galena, sphalerite, pyrite, and quartz. The ore shoots occur in the quartzites of probable Cambrian age, but the veins are small and nearly barren in the interbedded micaceous and arenaceous shales. The ores are oxidized to shallow depths, below which, in the more productive veins, the rich secondary silver sulph-antimonide minerals extend to the greatest depths to which the veins have been explored.

DEPOSITS IN GRANULAR IGNEOUS ROCKS.

VEINS.

Quartz veins were found in the granular igneous rocks in 7 of the 29 mining districts described in this report. In the Gold Butte district the veins are in granite and granite gneiss that are believed to be of pre-Cambrian age, but in the other six districts the veins are in intrusive granitic rocks that are believed to be either of late Mesozoic or early Tertiary age. In the Mizpah section of the Dolly Varden district and at Bald Mountain there are small bodies of leached mineralized quartz monzonite that somewhat resemble the disseminated " porphyry copper " deposits.

GOLD VEINS.

The narrow quartz veins in the pre-Cambrian granite of Gold Butte district are not strongly mineralized, and the zone of oxidation is shallow. Free-milling gold ore has been found in the upper parts of the veins, but at greater depth the gold is combined with the sparingly distributed sulphides, including pyrite and some chalcopyrite, galena, and sphalerite. In places fluorite occurs with the quartz, but it is nowhere abundant. The veins are either frozen or separated from the walls by thin selvage. Alteration of the wall rock by sericitization of the feldspars and chloritization of the biotite is not strong but is characteristic of the veins.

Veins that are valuable principally for gold and copper were found in the intrusive igneous rocks in the Delker, Dolly Varden (Mizpah section), Kinsley, Whitehorse, Troy, and Bald Mountain districts. As a rule the veins are narrow, rarely exceeding 2 feet in width and averaging about 10 inches. They consist of quartz carrying widely varying though nowhere large amounts of metallic minerals. In most of the veins the sulphide minerals are found at shallow depths and in some of them partly altered sulphides appear at the surface. Pyrite, chalcopyrite, and lesser amounts of galena and sphalerite are the characteristic minerals.

In the Mizpah section of the Dolly Varden district, Elko County, bismuthinite occurs in the ore and chalcopyrite is more abundant than pyrite. At Bald Mountain, in White Pine County, the quartz veins in the intrusive quartz monzonite carry stibnite and marcasite, besides cupriferous pyrite. Chemical tests show that traces of tellurium are present in this ore, but no telluride minerals were identified by the writer.

The solutions which mineralized the veins have also affected the adjacent wall rocks but only to a rather slight extent. The belts of altered rock on both sides of the vein are generally not wider than the vein itself, but at some places they are as much as 15 feet wide. The alteration consists of a sericitization of the feldspars and the ferromagnesian minerals, and the resulting rock is a soft white mass of quartz calcite and sericite, generally containing cubes of pyrite.

TUNGSTEN VEINS.

At the southeast side of Kern Mountains, in White Pine County, some quartz veins carrying hübnerite have been prospected. Several closely spaced northeastward-trending veins which cut the muscovite granite have been opened to shallow depths. They do not cut the overlying sedimentary rocks. The quartz contains (besides the dark reddish-brown crystals of hübnerite) fluorite, fragments of altered country rock, and crystals of mica.

DISSEMINATED DEPOSITS.

Two small bodies of highly altered quartz monzonite porphyry were seen during the summer of 1913—one at Mizpah, in Elko County, and the other at Bald Mountain, in White Pine County. The rock at both places is altered to a soft white mass of calcite, sericite, and quartz containing disseminated pyrite and possibly a small amount of chalcocite. The outcrop of the body at Mizpah is stained red, but the altered surface rock at Bald Mountain is only slightly iron-stained. Neither of these deposits have been opened to show bodies of workable ore, but they may be the upper leached part of a body similar to that worked at Ely.

DEPOSIT IN BIOTITE RHYOLITE TUFF.

The Atlanta ore body, in the Atlanta district, Lincoln County, is different from all the other ore bodies seen by the writer in 1913, in that it occurs in a fault breccia that is made up largely of fragments of biotite rhyolite tuff but that contains also fragments of limestone and quartzite. Free gold and some silver are the valuable constituents of the ore and are carried in limonitic material which partly cements the breccia, but pyrolusite is found in considerable amounts in the richer ore. The mineralizing solutions were evidently very siliceous, for the fault breccia and the rhyolite tuff have been considerably silicified, and the limestone near the main ore shoot east of the fault has also been silicified to some extent. In some of the hard ore a platy structure similar to that of the gold quartz of the " late Tertiary deposits " is common, and the quartz in this kind of material is seen to be replacing barite. No adularia, a common gangue mineral of the late Tertiary gold ores, was found in the Atlanta ore.

MINERALS IN THE ORE DEPOSITS.

For convenience of reference the minerals noted in the ores or closely associated with them are given below in alphabetic order, with brief notes on their occurrence. The list is obviously not complete, for many deposits in the region were not visited.

Amphibole.—A light-colored fibrous amphibolite of the variety tremolite $CaMg_3(SiO_3)_4$, identified from the Bi-Metallic and Victoria mines of the Dolly Varden district, associated with epidote and garnet in contact-metamorphic deposits and in the Ferber district with copper ores.

Anglesite$=PbSO_4$.—Forms thin envelopes around residual kernels of galena and is in turn surrounded by cerusite in the partly oxidized ores of the lead deposits. Particularly noticeable in the lead ores from the Ruby Valley or Smith Creek, Mud Springs, Spruce Mountain, Dolly Varden, Tecoma, Ravenswood, Hunter, and Bristol districts.

Argentite (silver glance)$=Ag_2S$.—A soft black mineral that cuts like wax. Identified in some of the rich ores of the Star mine, Cherry Creek district, and probably present in ores from other mines of that district. Was probably present in the rich ores of the Aurum, Kinsley, Spruce Mountain, and Silver Park districts.

Arsenopyrite$=FeAsS$.—Silvery white to gray metallic mineral found at the Rustler mine, Willow Creek district, associated with argentiferous tetrahedrite, and at the Siegel mine, Aurum district, with pyrite and galena.

Azurite$=Cu_3(OH)_2(CO_3)_2$.—A brilliant blue mineral that occurs as thin films on joints in the oxidized copper ores of this region; widely distributed, but nowhere found in large amounts. Found with the rich silver ores of Kinsley, Cherry Creek, Aurum, Atlanta, and Willow Creek districts.

Barite (heavy spar)$=BaSO_4$.—A white vitreous mineral, commonly occurring in thin plates. Found with quartz as a gangue mineral of the ores at the Dead Horse mine, Mud Springs district, and at Ferguson Springs, with limonitic copper ore. In the siliceous gold ore of the Atlanta mine, Atlanta district, microscopic examination shows silica replacing minute crystals of barite.

Bismuth$=Bi$.—Native bismuth, a silvery-white mineral, was found associated with bismuthinite, scheelite, epidote, rutile, and phlogopite in small lenslike masses of contact-metamorphic origin at the Hankins prospect, Valley View district.

Bismuthinite$=Bi_2S_3$.—A lead-gray fibrous metallic mineral; occurs with bismuth and scheelite in the Valley View district. Found also in the siliceous pyrite veins of the Mizpah section of the Dolly Varden district and in the hübnerite-fluorite veins at Regan, in the Kern Mountains.

Bornite (purple copper ore)$=Cu_6FeS_4$.—Reddish-brown metallic mineral found closely associated with chalcopyrite in copper ores of the Lincoln mine, Gold Butte district; Crescent prospect, Ruby Valley district; Banner Hill mine, Spruce Mountain district; Butte Group, Dolly Varden district; and in the southern prospects on the Kinsley consolidated ground, Kinsley district.

Carnotite.—A canary-yellow ocherous mineral containing uranium and vanadium. Found coating joints in ore and brecciated rhyolite tuff at the Atlanta mine, Atlanta district.

Cerargyrite (horn silver)$=AgCl$.—A soft grayish-green or brown waxy mineral. Occurs in considerable amounts at the Silver Park mine, Atlanta district, and in smaller amounts in ores from the Morning Star mine, Kinsley district, and Jackrabbit mine, Bristol district. Probably also present in minute quantities, in the highly argentiferous lead carbonate ores of this region.

Cerusite (lead carbonate)=$PbCO_3$.—White to gray vitreous mineral, commonly ocherous. Occurs as crystalline masses and in the form of sand in the surface ores of most of the lead deposits. Results from the surface alteration of galena, and may carry silver. Is one of the most valuable ore minerals of the Spruce Mountain, Dolly Varden, Hunter, Ward, Red Hills, and Bristol districts.

Chalcocite (copper glance)=Cu_2S.—A rather soft, black metallic mineral. In this region found as an alteration product of chalcopyrite and bornite, coating or completely replacing the older sulphides. Generally coated with copper carbonates. Occurs in appreciable amounts in copper deposits of the Spruce Mountain, Dolly Varden, and Kinsley districts, and is common in the copper ores of the camps.

Chalcopyrite (copper pyrites)=$CuFeS_2$.—A bright-yellow metallic mineral found in varying amounts in all the copper deposits described in this report.

Chrysocolla=$CuSiO_3+2H_2O$.—A green to blue incrustation on joints in the surface ores of copper deposits. Often mistaken for copper carbonate, but can readily be determined, as it does not effervesce with acid. Is more common than the carbonates in very siliceous deposits, such as the Victoria and Eugene, at Dolly Varden, and in the quartz veins of Cherry Creek.

Copper pitch ore.—Though not a distinct mineral species, this material is so commonly found in the surface ores of the copper deposits of eastern Nevada that it should be recognized. Lindgren[1] describes copper pitch ore from Clifton-Morenci, Ariz., as " a dark-brown to black substance, sometimes dull, but generally with glassy to resinous luster; hardness, about 4; streak, dark brown. * * * Most of these copper pitch ores * * * probably represent a series of closely related compounds." Chemical analysis shows that copper pitch ore is chiefly a combination of copper, iron, and manganese oxides, but may contain other oxides in varying proportions. The copper pitch ores of eastern Nevada are usually of some shade of brown, but from the Luray and Bristol districts a nearly black variety was obtained.

Cuprite (red copper oxide)=Cu_2O.—Dark-red masses of cuprite were noted in the surface ores from the Gold Butte and Bald Mountain copper deposits, and it is believed that this mineral is present in much of the limonitic copper ores of the Bristol district.

Enargite=Cu_3AsS_4.—A black, brittle mineral, sometimes mistaken for gray copper. In the Taylor and Kern districts a silver-bearing enargite was found with galena and sphalerite in siliceous ores.

Epidote=$Ca_2Al_2(AlOH)(SiO_4)_3$.—A peculiar yellowish-green mineral associated with garnet and amphibole in the contact-metamorphic copper deposits at Spruce Mountain, Dolly Varden, Kinsley, and Aurum districts. A light cream-colored variety, clinozoisite, was found in the bismuth-tungsten deposits of the Valley View distrct.

Fluorite=CaF_2.—A medium soft vitreous mineral, usually found in cubical crystals in cavities, as in the quartz-pyrite veins of the Gold Butte district and in the quartz-hübnerite veins at Regan, in the Kern Mountains. Occurs in microscopic crystals with pyroxene, pyrite, and sphalerite in a bed of metamorphosed shale in the Patterson district.

Galena=PbS.—A soft lead-gray mineral with metallic luster and cubical cleavage. Was found in practically every district visited. Occurs as replace-

---

[1] Lindgren, Waldemar, The copper deposits of the Clifton-Morenci district, Ariz.: U. S. Geol. Survey Prof. Paper 43, pp. 114–115, 1905.

ment deposits in limestone, in quartz veins with pyrite and sphalerite, and in contact-metamorphic deposits with chalcopyrite, pyrite, and sphalerite. In this region it is usually argentiferous; in the Ravenswood district it is antimonial. In many of the districts visited galena is found only as residual kernels in the cerusite ores, but it is the chief metallic mineral of the Cuba mine, in the Granite district, and with pyrite, sphalerite, and tetrahedrite forms the ore of the lower levels of the Star mine, at Cherry Creek.

Garnet.—A variable calcium-magnesium-iron aluminum silicate that in eastern Nevada is usually reddish brown or yellow brown. It occurs with epidote, amphibole, and pyroxene in the zones of contact-metamorphosed rocks in the Valley View, Spruce Mountain, Dolly Varden, and Aurum districts. Copper minerals are generally associated with the zones of contact alteration.

Gold.—A soft, brilliant yellow metal occurring free and in conspicuously visible grains in the ore of the Melbourn and other calcite veins of the Willow Creek district, where it is always associated with green talc. Occurs free but usually in very minute grains at the Butte group, Dolly Varden district; Atlanta mine, Atlanta district; in the siliceous veins of Egan Canyon, Cherry Creek district; and in the Granite district. Gold has been recovered from placer gravels at Bald Mountain and Egan Canyon.

Hübnerite.—Supposed to be a nearly pure manganese tungstate. Occurs in large quartz veins at Regan, at the south end of the Kern Mountains. The veins are found in a coarse-grained biotite-muscovite granite. Fluorite, bismuthinite, and triplite are found with the hübnerite.

Limonite=$Fe_2(OH)_6Fe_2O_3$.—Earthy yellow to brown material found in the surface ores of every district described in this report. Particularly abundant at the Lincoln mine, Gold Butte district, and in many of the mines of the Bristol district. At Spruce Mountain plumbojarosite is intimately mixed with the limonite.

Marcasite (white iron pyrites)=FeS_2.—A light-yellow metallic mineral found in small amounts with the pyritic quartz veins of the Bald Mountain district.

Malachite=$Cu_2(OH)_2CO_3$.—Forms bright-green masses and crusts in the surface ores of the copper deposits in eastern Nevada. There are no large deposits of this mineral in eastern Nevada, though it is widely distributed in small amounts.

Molybdenite=MoS_2.—Small bluish-gray metallic scales were noted in siliceous pyritic ore of the McMurry prospect at Cherry Creek.

Phlogopite (bronze mica).—Small flakes of red-brown mica occur with garnet, epidote, amphibole, and pyroxene in the contact-metamorphic copper-lead deposits of the Ruby Valley district and with the bismuth and tungsten bearing lenses of contact-altered limestone of the Valley View district.

Plumbojarosite.—A complex lead-iron hydrous sulphate. Occurs as hexagonal flakes in the limonitic and cerusitic ores of the Tecoma and Spruce Mountain districts. May occur also in the Bristol lead ores, though it was not definitely determined.

Proustite (light ruby silver)=Ag_3AsS_3.—Small crystals of a reddish-gray mineral with scarlet streak were found in the enriched silver ores of the Star mine at Cherry Creek and of the Silver Park mine at Atlanta.

Pyrite=FeS_2.—A brass-yellow metallic mineral; crystallizes in cubes. Widely distributed with copper minerals in contact-metamorphic replacement and vein deposits but not so abundant in the replacement deposits carrying lead. Also commonly seen adjacent to quartz veins in softened igneous rocks that have been altered by the solutions that deposited the ores.

Pyrolusite=MnO_2.—A soft black mineral associated with the limonitic lead and copper ores of the Bristol and Bald Mountain districts and with limonite in the free gold ores of the Atlanta mine, Atlanta district. At the Siegel mine, in the Aurum district, large bodies of argentiferous pyrolusite were found in the upper levels and smaller bodies of similar ores were found at the Lucky Deposit mine, in the same district. In the Day (Jack Rabbit) mine, Bristol district, some large stopes of limestone replaced by argentiferous pyrolusite were mined in the lower leved. Small quantities of this mineral are generally present in surface ores of the copper deposits.

Pyromorphite=$Pb_5Cl(PO_4)_3$.—A peculiar greenish lead phosphate, an alteration product of galena in the veins, occurs in small patches at Ravenswood.

Pyroxene.—A fibrous iron-magnesium-aluminum silicate with a greenish-white color and vitreous to resinous luster, associated with other lime silicate minerals in contact-metamorphic deposits. Variety diopside, determined from ores collected at Smith Creek and in the Spruce Mountain and Patterson districts.

Scheelite=$CaWO_4$.—Occurs as cream-white crystals in lenses of contact-metamorphic origin associated with phlogopite, rutile, and epidote at the Hankins prospect, Valley View district.

Silver=Ag.—Native silver was found in the rich ores of the Rustler mine, Willow Creek district, associated with copper pitch ore that had been formed by the alteration of arsenopyrite and argentiferous tetrahedrite.

Smithsonite (dry bone)=$ZnCO_3$.—A yellowish-white earthy mineral associated with the cerusite ores of the Warm Creek, Mud Springs, Tecoma, Aurum, and Bristol districts.

Sphalerite=ZnS.—A yellowish-brown to black resinous mineral, with white streak, found in the pyrite-galena-quartz veins of Gold Butte, Troy, and Kern districts. In the district last named it is accompanied by enargite. It occurs with galena and enargite at Taylor and with galena and tetrahedrite at Cherry Creek. At Smith Creek it occurs in contact-metamorphic deposits with either galena or chalcopyrite, and at Patterson with pyrite, pyroxene, and fluorite.

Stibnite=Sb_2S_3.—A bright lead-gray to black fibrous mineral occurring in minute needles in the pyritic gold quartz veins of the Bald Mountain district, and having vivianite as its alteration product in small replacements of limestone.

Tetrahedrite (gray copper)=$Cu_8Sb_2S_7$.—A fine-grained dark-gray mineral, which in this region is generally argentiferous; determined in ores from the Ravenswood, Willow Creek, and Bald Mountain districts. At the Star mine in Cherry Creek it was found in considerable amounts, both as a primary and secondary deposition.

DEPOSITS AND MINES.

CLARK COUNTY.

GOLD BUTTE MINING DISTRICT.

LOCATION AND ACCESSIBILITY.

The Gold Butte district (No. 1, fig. 1, p. 18) covers the south end of the Virgin Range, lying north of the big bend of Colorado River, west of Virgin River, east of Grand Wash, and south of Bitter Spring Pass. The mines are near the crest of the range, and the principal camp at Voigt well is about 28 miles southeast of St. Thomas. (See fig. 2.) That old Mormon settlement at the junction of the Virgin and Muddy rivers is now served by a branch of the San Pedro, Los Angeles & Salt Lake Railroad, which leave the main line at Moapa.

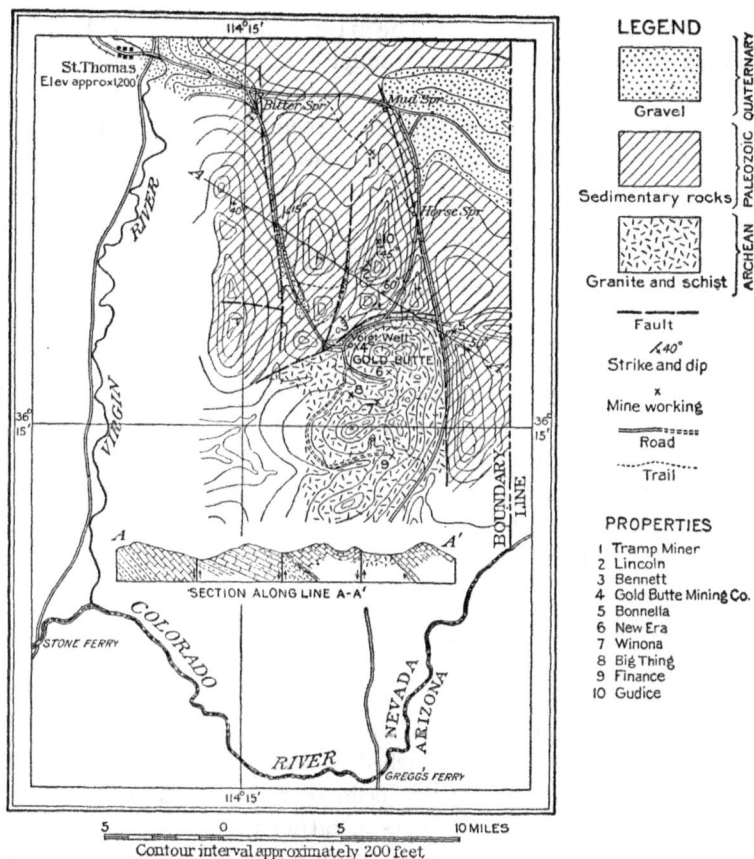

FIGURE 2.—Sketch map of the Gold Butte district, Clark County, Nev., showing approximate topography, major geologic relations, and the positions of the mines, roads, and watering places.

TOPOGRAPHY.

The Virgin Range is composed of a series of more or less parallel northward-trending ridges north of Gold Butte, but south of that latitude it is an irregular, deeply dissected mountain mass.

Its western face, as seen from St. Thomas, rises from the valley of the Virgin, whose elevation is about 1,200 feet above sea level, in a series of terraces to Gold Butte, whose elevation is about 4,500 feet above sea level. The eastern wall, on the contrary, rises abruptly from Grand Wash in a bold, steep slope to a northward-trending ridge that is separated from the main mass of the mountains by a deep gorgelike canyon, which is followed by the road from Horse Springs to Greggs Ferry. (See fig. 2.)

The more or less parallel northward-trending ridges shown in the upper part of figure 2 have cliff-like western faces and more gentle though still steep eastward slopes. They are separated by moderately deep, straight canyons. The arrangement of these ridges brings out in a remarkably striking manner the faulted structure of that part of the area shown in the cross section.

The narrow well-marked valley between the regular ridges on the north and the irregular mountain mass to the south is a distinct feature of the topography. This depression follows a fault zone whose northern wall forms about $1\frac{1}{2}$ miles north of Voigt well a prominent narrow ridge of nearly perpendicular eastward-striking sedimentary beds. (See Pl. II, *B*.)

GEOLOGY.

The oldest rocks in the Gold Butte district are pre-Cambrian granite, gneiss, and schist, which apparently underlie most of the area south of and including Gold Butte. The sedimentary rocks forming the eastern and northern part of the area shown in figure 2 are of Paleozoic age. Partly consolidated Quaternary gravels cover a considerable portion of the Grand Wash and are also seen along the Muddy and Virgin valleys.

PRE-CAMBRIAN ROCKS.

Biotite granite gneiss is the prevailing rock of pre-Cambrian age in the area south of Gold Butte. It is for the most part a light to dark gray medium-grained rock, which does not everywhere show gneissic banding yet in some places has been so metamorphosed as to have become a mica schist. It weathers in rather ragged blocks. The rock is composed of quartz, orthoclase, microcline, and some oligoclase with more or less abundant dark olive-green biotite. In the schistose facies the biotite is very abundant, but some variations of the gneiss contain only small amounts of the dark mica. The feldspars in most of the hand specimens collected are somewhat sericitized, and the biotite in one thin section is chloritized.

Gold Butte is a roughly elliptical mass of coarse-grained porphyritic granite about 4 miles long from east to west by 2 miles from north to south. This rock has been intruded into the gneiss, but as it does not appear to have cut any of the sedimentary rocks it is thought to have been intruded in pre-Cambrian time, though its age is not certain. This view is somewhat supported by the fact

A. NORTH SIDE OF GOLD BUTTE.

View westward along the depression between the granite and the sedimentary rocks.

Fault Ridge

B. VIEW EASTWARD UP BITTER SPRING WASH, ABOUT 2 MILES WEST OF THE SPRING.

C. SOUTHWEST SIDE OF GOLD BUTTE.

Looking N. 40° E., showing the sheeting and weathering of the coarse-grained granite.

that the mass is broken by a closely spaced series of joints roughly parallel to the major northward-trending fault lines in the region. (See Pl. II, *C.*)

The rock is pinkish gray when fresh, weathers dull brown in rounded forms, well shown in the illustration, and by disintegration has given rise to an immense amount of coarse arkosic sand, which fills the valleys north, west, and south of the peak. Over half the bulk of the rock in the hand specimens is feldspar; quartz is next in abundance and is followed by biotite and some hornblende. Examination of the thin sections under the microscope shows that the feldspars are of several varieties. Large phenocrysts of microperthite inclose small fragments of quartz and orthoclase. Microperthite, orthoclase, microcline, and a small amount of oligoclase are present in the coarsely granular groundmass. The biotite is dark olive-green and shows strong pleochroism. Green hornblende is not abundant and is usually closely associated with the biotite. Zircon and apatite are common accessory minerals.

Dikes of fine-grained nearly white aplite are seen in places in both the granite and schist. Pegmatitic masses consisting of feldspar, quartz, and biotite are abundant in the granite, particularly on the eastern peak of Gold Butte.

PALEOZOIC ROCKS.

The area mapped as sedimentary rocks on figure 2 is underlain by limestones and sandstones of Paleozoic age. At the base of this series west of the Bonnella mine (No. 5, fig. 2) there is from 75 to 90 feet of light-colored sandy quartzite. At this locality there has been some strike faulting along the contact, and small pink garnets are developed in the quartzite. Upon the quartzite lies 100 to 200 feet of reddish-purple shaly limestones, overlain in the vicinity of the Bonnella mine by light-gray crystalline limestones, which occur in beds from 6 inches to 2 feet thick and weather dull red. Interbedded with the limestone are some red and yellow lime shales, which are considered the equivalents of the lower part of the Cambrian Tonto group of the Grand Canyon section described by Gilbert.[1]

The beds dip east at steep angles on the contact, and at about 45° in the vicinity of the Bonnella prospect. They form a huge hogback that extends for several miles southward from the mine.

The northward-trending ridges north of Gold Butte, which are separated by faults, are composed of limestones dipping 45° to 60° E. (See fig. 2.) The main ridge, on which most of the prospects

---

[1] Gilbert, G. K., U. S. Geog. and Geol. Surveys W. 100th Mer. Rept., vol. 3, pp. 162–163, 184–185, 1875.

are located, is composed of thin-bedded, pinkish-gray crystalline limestones with much dark chert in some beds. There are branching twiglike bodies of calcite in certain beds, which somewhat resemble fossils, though no identifiable species were found in the small collection of the material brought in for study. Edwin Kirk, of this Survey, considers these forms pre-Carboniferous. Gilbert[1] describes similar markings on what he called lower Silurian limestone exposed at the mouth of the Grand Canyon. Along the west base of the ridge, west of the Lincoln mine (No. 2, fig. 2), there is a belt of dark-colored cherty limestone which appears to be much fractured. What is thought to be the bedding dips about 80° E. This belt is bounded on the west by a long, narrow outcrop of granitic rock which was at first thought to be intrusive but which is now believed to be a faulted segment of the pre-Cambrian.

The two ridges on either side of the road from Bitter Springs to Voigt well are composed of light pinkish-gray limestone, which overlies a persistent 50-foot bed of cross-bedded dark-red sandstone, below which there is a considerable thickness of thin-bedded red and green calcareous shale and sandstone. In these beds, particularly the red sandy portion exposed for a mile south of Bitter Springs, gypsum is seen in small veinlets parallel to and cutting across the bedding. On some of the calcareous sandy shales ripple marks were noted. In a low saddle about 5 miles northwest of Voigt well a few fossils were collected from the upper limestone, which G. H. Girty, of the Survey, has examined. He found " Crinoid segments, large and small, together with fragments of a shell, which appears to belong to the genus *Composita.* * * * It seems probable that this lot is of Carboniferous age, and the lithology suggests Mississippian." The formations in these ridges are thought to represent the lower or middle part of the Redwall limestone, which is exposed at the mouth of the Grand Canyon and has been described by Gilbert.[2]

QUATERNARY DEPOSITS.

Bitter Spring Wash from Virgin River eastward nearly to the spring is deeply intrenched in partly consolidated, nearly horizontal gravels (see Pl. II, *A*), which rest unconformably upon other sedimentary rocks. The gravels include subangular cobbles of sandstone and limestone in a matrix of calcareous sandy cement and many lenses of sand. The bedding, though in most places distinct, is irregular. These gravels appear to have been deposited by moving water, rather than in lakes, and strongly resemble the Gila conglomerate of southern Arizona.

---

[1] Gilbert, G. K., op. cit., pp. 184–185.   [2] Idem, p. 178.

Spurr[1] thought these gravels might be of Pliocene age. Lee[2] has described similar gravels at the mouth of Virgin River, which he named the Temple Bar conglomerate, noting their resemblance to the Gila.[3] It is his opinion that they were formed in Pleistocene time, largely by moving water but partly in a lake formed behind a dam of basalt 800 feet thick, which was thrown across the Colorado River valley, near the mouth of Williams River. The gravels along Virgin River are very similar to the gravel fill of Grand Wash east of the Virgin Range. A small part of these gravels is shown in the upper right corner of figure 2. In Grand Wash basalt flows are intercalated with the gravels, as noted by Marvine.[4]

BASALT.

A considerable area southwest of Gold Butte is underlain by a dark greenish-black vesicular flow rock which is thought to be basalt. The rock was examined in the field at only one small isolated exposure and no specimens were obtained for study.

STRUCTURE.

According to Marvine[5] the northern part of the Virgin Range has an anticlinal structure. In the part of the range in the vicinity of Gold Butte, as will be seen from an inspection of the section on figure 2, there are two northward-trending faults, of considerable throw, which have divided the range into monoclinal ridges having an eastward dip. Besides these major northward-trending faults there are a number of less noticeable northward-trending breaks. The strike fault between the pre-Cambrian mass and the hogback east of Gold Butte is probably not of great magnitude.

This primary northerly trend is complicated by at least two faults with easterly strikes. Along the southern of these faults there was considerable displacement, as the pre-Cambrian gneiss is brought against Mississippian sedimentary rocks. At the contact the silicified limestones dip 80° N. for a short distance and then swing into the general easterly dip of the main ridge. (See Pl. II, *B*.) Along the eastward-trending fault west of the road between Bitter Spring and Gold Butte the displacement was apparently not great. The dip of the strata north of the fault is to the east at about 45°. For a few

[1] Spurr, J. E., Descriptive geology of Nevada south of the fortieth parallel and adjacent portions of California: U. S. Geol. Survey Bull. 208, pp. 131–132, 1903.

[2] Lee, W. T., Geologic reconnaissance of a part of western Arizona: U. S. Geol. Survey Bull. 352, pp. 17–18, 1908.

[3] Idem, pp. 58, 63–64.

[4] Marvine, A. R., U. S. Geog. and Geol. Surveys W. 100th Mer. Rept., vol. 3, pp. 196–198, 1875.

[5] Idem, p. 194.

hundred feet south of the fault the pinkish-gray limestones dip 20° S., but they flatten to nearly horizontal in less than half a mile.

ORE DEPOSITS.

HISTORY AND PRODUCTION.

So far as could be learned the veins in the Gold Butte region were discovered about 1907 and were worked in the following two years. They were discovered by Messrs. Bonnella, Burgess, Syphus, and Gentry, all of Muddy Valley. In 1908 there was one rather mild "excitement" in the region, during which the townsite of Gold Butte was laid out in the flat west of Voigt well. In the fall of 1913 seven men were working in the district—two at prospects near Greggs Ferry, on Colorado River, and the remainder in the vicinity of Gold Butte.

So far as known the only ore shipped from this district consisted of two small lots of oxidized copper ore taken from shallow workings in the Tramp Miner property and the Lincoln mine (Nos. 1 and 2, fig. 2).

CHARACTER AND OCCURRENCE OF THE ORES.

Two distinct types of ore deposits are found in the Gold Butte district—replacement deposits in limestone and quartz veins in the pre-Cambrian gneiss and granite.

Small, irregular replacement deposits in limestone, most of them along minor fractures, have furnished the only ores shipped from the district. The ores consist of hematite, limonite, and lesser amounts of oxidized copper minerals. At the Lincoln mine a small amount of chalcocite, secondary after chalcopyrite and bornite, was found. In all the prospects in deposits of this type the ores seem to be found most often at or near the surface. At the Lincoln, Bennett, and Bonnella mines tunnels 25 to 60 feet underground showed only slight mineralization, though the porous beds or fault breccias suitable for ore deposition were followed by the workings. It seems probable that these ores were formed by downward-moving cold waters not heavily charged with mineral and that all their metallic burden was deposited near the surface. They probably represent concentration along particularly favorable channels.

The age of these deposits is problematic, yet there is a suggestion that they may have been formed during that late period of geologic time in which a large number of copper deposits were formed in the "Red Beds" of the Southwest.

The veins in granite and granite gneiss strike prevailingly north-northeast and have steep dips where they show any inclination from the vertical. Those seen are small and not strongly mineralized. Milky-white quartz is practically the only gangue mineral, though fluorite was noted in the New Era vein. The veins have a strong tendency to branch and unite along the strike, but the branches rarely leave the lode for any great distance. All the veins show a central barren quartz filling, with the sulphides sparingly distributed next the walls. Pyrite is by far the most important sulphide, though chalcopyrite, galena, and sphalerite are seen. The outcrops are not strongly iron stained, as the metallic content of the veins is small, and for long distances barren white quartz alone is seen on the strike of even the most important veins.

The country rock adjacent to the quartz is slightly altered. The alteration of the coarse-grained porphyritic granite is more intense than the granite gneiss, though it is of the same character in both types of rock. The biotite is altered to chlorite, and the feldspars, particularly the plagioclase and microcline, are changed to sericite.

At the surface of the more heavily mineralized portions of these veins a little free gold has been found. It seems doubtful whether they will ever prove to be of much value because of the scarcity of sulphides in so much of the quartz and because of the low grade of the ore found below the water level, which is at most places not far from the surface.

DEPOSITS IN LIMESTONE.

Tramp Miner property.—About 3 miles south-southwest of Mud Spring, on the west side of the main ridge north of Gold Butte, there is a small group of claims known as the Tramp Miner property (No. 1, fig. 2). This group is controlled by B. W. Whitmore, of Bunkerville, but was last worked by Bishop Whitehead, of Overton. The development consists of a number of inclined shafts following the ore zone, the deepest about 125 feet long, with short levels at 50 and 100 feet.

The deposit occurs in limestones that dip east at medium angles along a more or less open bedding plane, near a zone of fracturing that is about parallel to the bedding. The ores are largely limonitic but contain irregular pockets and stringers of cuprite and malachite. It is said that the single small shipment of sorted ore from this property carried about 24 per cent copper to the ton.

Gudice claims.—The Gudice claims (No. 10, fig. 2) are located on the summit of the main ridge about half a mile northeast of the Lincoln mine. They were not visited, but it is said that the little development work which has been done shows iron oxide ore carrying some copper carbonates occurring in limestones, as at the Tramp and Lincoln mines.

Lincoln mine.—The Lincoln mine (No. 2, fig. 2) is developed by a shallow incline and two tunnels located about halfway between the base and summit of the steep west face of the main ridge 3½ miles north of Voigt well. There were three claims in the group, originally controlled by H. W. Burgess. It is understood that no work has been done on the property since 1911.

The incline shaft, 50 feet long, dips to the east at 25°, following a bedding plane of limestone along which there are 8 to 10 inches of limonite. At the bottom of the incline a tight fault striking N. 10° W. and dipping 45° W. cuts off the ore. The direction of movement along this slip could not be ascertained, though it seems probable from the nature of the faults in this region that the west side must have dropped with respect to the beds east of the fault.

The tunnels are 150 feet north of the collar of the incline, and the upper drift tunnel is 50 feet below it. This drift runs S. 35° W. for about 40 feet in light pinkish-gray crystalline limestone below a bed of pearl-gray limestone with abundant white chert. The ores occur in a narrow zone below the chert bed in small open watercourses. Limonite and occasionally small kidneys of green copper carbonate are irregularly distributed through the coarsely crystalline white calcite. In a piece of rich heavy dark carbonate ore some chalcocite was seen surrounding minute remnants of chalcopyrite and bornite. Fourteen tons of sorted high-grade copper ore was shipped from a pocket at the mouth of this tunnel.

The lower tunnel is about 75 feet below that just described. A crosscut 175 feet long runs S. 60° E., cutting, near the end, what appears to be the same porous limestone bed that was shown in the upper work. A few small stringers of limonite are exposed in the 75-foot drift south along the zone, but no copper minerals were noted.

Bennett mine.—The Bennett or Quartz King group (No. 3, fig. 2), belonging to Harvey Bennett, of Overton, covers a low hill a mile northwest of Voigt well. (See fig. 3.) The limestones of this hill strike north and dip east at steep angles. They are cut by a series of faults with small displacement which trend from north through northeast to east. On the west side of the hill there is a fracture striking N. 35° E. and dipping 50° ESE., along which there is a heavy clay gouge. The limestones for 15 feet west of this fault are replaced by massive red limonite and for 10 feet farther west by yellow limonite in gradually diminishing amounts. Near the south end of this body there is an inclined shaft said to be 100 feet deep, following the hanging wall of the ore. At the surface, particularly near the hanging wall, there are small veinlets and pockets of copper carbonate ore. The replacement is not complete, and even where the ore appears to be solid limonite there is a considerable amount of lime carbonate.

The two tunnels on the north side of the hill (see fig. 3) are driven along faults. The lower tunnel runs south for 100 feet along a

2-foot zone of limestone fault breccia, which is slightly limonitic. The movement along this fault was practically horizontal. The upper tunnel for the first 50 feet runs S. 10° W. along the east side of a limestone breccia, which is partly replaced by iron, and then turns S. 35° W. for 100 feet, following under a clay gouge in limestone breccia which is only slightly iron stained.

The long crosscut tunnel from the gulch southeast of the hill runs N. 18° E. for 200 feet parallel to the bedding of the limestones, which dips 40° E. It then turns due north along a barren vertical fault breccia. One hundred and twenty-five feet from the mouth of

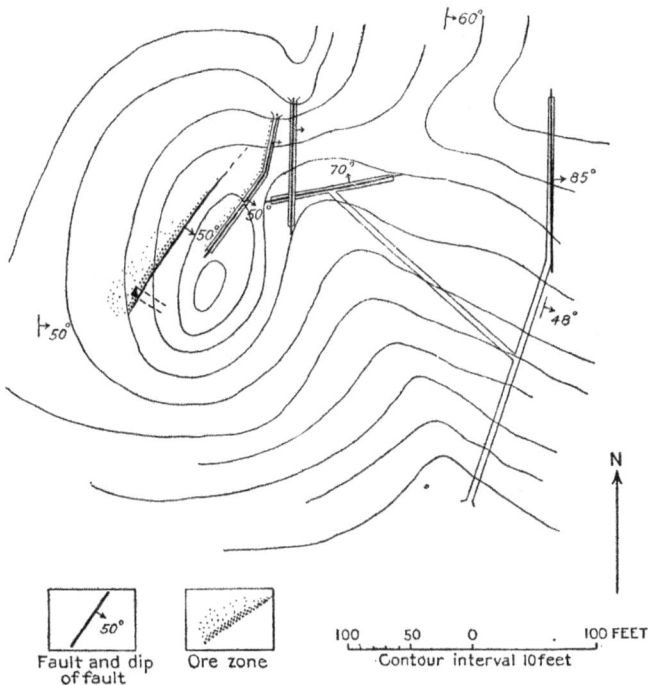

FIGURE 3.—Sketch map showing approximately the topography and relations of the work at the Bennett mine, Gold Butte district, Clark County, Nev.

the tunnel a crosscut is turned N. 50° W. and has been driven 200 feet to a fracture trending N. 80° E. and dipping steeply to the north, which has been drifted upon 50 feet each way. This barren fracture is tight, and the movement along it formed no breccia.

Bonnella properties.—The four Bonnella claims (No. 5, fig. 2) are about 6 miles east of Voigt well and half a mile east of the contact of the granite and the sedimentary rocks. The claims lie along the west side of the summit of the hogback formed by the upturned Paleozoic limestones. The principal development is in two short tunnels near the deep canyon which is cut through the hogback on a line with the major eastward-trending fault north of Gold Butte. There are some shallow workings half a mile south of the road near the summit. All this work is along a bed of massive light blue-gray limestone containing small irregular replacement lenses of limonite, which at a few places on the surface carry small amounts of copper silicate. These replacement bodies are associated with fractures that strike northeast or east, cutting the bedding, which strikes N. 10° E. and dips 45° E.

DEPOSITS IN GRANITE AND GNEISS.

Gold Butte Mining Co.—The Gold Butte Mining Co. formerly controlled a group of claims covering the west end of Gold Butte (No. 4, fig. 2), on which several small quartz veins were found. These, as seen on the surface, have a north to northeast course and are nearly vertical. The company drove a long crosscut tunnel southeastward under the butte from a point about half a mile east-north-east of Voigt well. This tunnel is about 800 feet long and for the first 440 feet trends S. 17° E. along a slip plane which dips 45° W. and is marked by a thin gouge, at which the tunnel turns S. 42° W., continuing for 135 feet. The coarse porphyritic granite is bleached and softened for 4 to 8 inches on either side of the fracture. Fifty feet from the turn two subparallel fissures about 19 inches apart strike N. 51° E. and dip 50° NW. Between these fissures the granite is so soft that it has "run" into the tunnel, leaving a cave about 15 feet high above the roof. The granite for 15 feet on the side of the fissures is sericitized and contains some pyrite, which is more abundant near the fissures. The last 160 feet of the tunnel is driven on a tight slip striking N. 52° W. and dipping 85° SW., along which the granite is scarcely altered. Considerable water issues from this fracture.

New Era group.—The New Era group (No. 6, fig. 2) of eight claims, of which three are on the main vein, is in the flat southeast of Gold Butte, not far north of the granite gneiss contact. It is the property of W. H. Liston and Scott Allen. The main work is a 118-foot shaft, now under water below a depth of 40 feet, at which there is a 12-horsepower gasoline hoist. There are numerous shallow shafts and tunnels along the vein, which can be traced for a mile. It strikes N. 85° E. and is nearly vertical in most of the places where it was seen. The vein ranges in width from 2 to 4 inches, and here and there splits into stringers around small masses of the coarse porphyritic granite country rock. It is at many places frozen to both walls, but at others a thin clay selvage permits the quartz to separate from the granite. The granite along the vein is altered in a zone measuring 2 to 8 inches. In this zone the feldspars are altered to sericite and a little calcite and the biotite to chlorite. The vein consists for the most part of white quartz, though one small aggregate of greenish-tinged fluorite was noted. The quartz is well crystallized, and vugs lined with small clear quartz crystals are common. Metallic minerals are sparingly and irregularly distributed, usually near the walls of the vein. Pyrite, largely altered to limonite above the water level, is the most common mineral, though a little galena is occasionally found.

Winona group.—The Winona group of two claims (No. 7, fig. 2), belonging to Levi Syphus and Harry Gentry, of St. Thomas, is about 2 miles south of the New Era and 6 miles southeast of Voigt well. A white quartz vein, from 2 to 10 inches wide, which strikes N. 65° E., has been followed south for 200 feet by a tunnel. The vein is frozen to the granite gneiss walls. Pyrite and occasionally chalcopyrite are sparingly scattered through the quartz near the walls.

Big Thing tunnel.—The Big Thing tunnel (No. 8, fig. 2) was driven by Olli Rosson. It runs S. 65° W. for 300 feet along a white quartz vein in granite gneiss. At some places altered gneiss for a width of 4 feet is cut by small stringers of quartz, and at other places the vein narrows to a single band of quartz about 8 inches wide. Chalcopyrite, galena, and sphalerite are here and there present in the border zone of quartz, which, however, usually carries only pyrite.

Finance group.—The Finance group of four claims (No. 9, fig. 2) is owned by Messrs. Syphus and Gentry, of St. Thomas. The principal development is a whim shaft, said to be 140 feet deep, in which water stands at a depth of about 15 feet. This shaft is sunk on a lodelike body of altered granite gneiss that is cut by numerous veinlets of white quartz carrying some sulphides, which strikes N. 35° E. and is nearly vertical. At some places the zone is 4 feet wide; at others it narrows to a single quartz vein 10 to 12 inches wide.

The granite gneiss is chloritized along the vein, but at a distance of a foot from the quartz it is bleached and sericitized. Thin clay partings are seen at some places between the walls and the quartz, which has here and there been crushed. The sulphides occur at the outer edge of the quartz and are not abundant. Pyrite and chalcopyrite are the most noticeable metallic minerals in the partly oxidized ore seen on the dump, though some galena and a little sphalerite were noted. The oxidized iron-stained and copper-stained ore above the water level is said to assay from $18 to $20 a ton in gold and silver.

Other prospects.—It is reported that two men are working a quartz vein in granite gneiss a short distance north of Greggs Ferry. This property was not visited. It is said that considerable development work has been done and that the ore treated in a small stamp mill at Colorado River yielded a small amount of bullion.

ELKO COUNTY.

ORDER OF DESCRIPTION.

In Elko County, which was part of Lander County prior to 1870, there are at least 20 mining districts. During the reconnaissance made in 1913 the writer somewhat hurriedly covered the region south of the line of the Southern Pacific and east of the Ruby Mountains. Approximately correct positions of the 13 camps visited are shown on figure 1 and Plate I.

In the following pages the districts in the same range or group of hills are described together, the Warm Creek, Ruby Valley, and Valley View districts, on the east side of Ruby Range, being described first, and the Mud Springs and Delker districts in the Ruby Hills and Franklin Buttes following. The mines at Spruce Mountain, at the south end of the Peoquop Range, are next discussed. The Dolly Varden (Mizpah) district, in the Wachoe Mountains, and the Kinsley district, in the northern part of the Antelope Mountains, are next described; next follow notes on the Luray, Ferguson Spring, White Horse, and Ferber districts, in the Toano Range. Lastly, the Tecoma district, at the extreme eastern border of the State, in the Goose Creek Hills, north of the Southern Pacific line, is described.

RUBY RANGE.

LOCATION AND GENERAL FEATURES.

The Ruby Range, known to the geologists of the Fortieth Parallel Survey as the Humboldt Range, is in southwestern Elko County. The map of Nevada issued by the General Land Office in 1908 shows the main range as the Ruby Mountains and the smaller group of mountains east of Clover Valley as the East Humboldt Range. The main range is about 80 miles long, extending from latitude 40° N. to 41° N., and from longitude 115° W. to 115° 45′ W. It trends about N. 20° E.

Elko, the county seat, about 25 miles west of the north end of the range, is the principal supply point for the settlements in South Fork Valley east of the mountains, though Halleck is the shipping point. Wells, on the Southern Pacific, east of the north end of the range, is the shipping and supply point for most of the population of Ruby and Clover valleys, but of late years Currie, on the Nevada Northern Railway, has taken some of the produce. Elko naturally is not entirely abandoned as a base, particularly for the ranchers in Clover Valley. Prosperous ranches in the valley on both sides of the range are located at the mouths of the many streams which rise in the mountains. Ruby and Clover valleys, on the east side of the range, are particularly fertile, and a considerable area north of Franklin Lake (see Pl. III) is probably available for dry farming, as water stands near the surface.

TOPOGRAPHY AND DRAINAGE.

The Ruby Range (see Pl. III) is narrow and very rugged and includes some of the highest peaks in the State of Nevada. Its height and its ruggedness are accentuated by the fact that the valleys on both sides are broad, flat, and lie at a rather low elevation. There are three passes by which the range can be easily crossed. Sacred Pass, the northernmost, is low and is traversed by an excellent road. Harrison Pass, formerly known as Fremont Pass, which crosses the center of the range west of the north end of Ruby Lake, is comparatively low. The road through this pass has recently been regraded by the United States Forest Service, so that it now can be easily traversed by heavy wagons. The old Overland Stage Road through Ruby Pass, formerly known as Hastings Pass, at the south end of Ruby Lake in White Pine County, has not been much traveled of late years, and the heavy rains in August, 1913, are said to have washed out most of the old grade.

The summit of the range is nearer the eastern than the western mountain front, and the east face of the mountains presents a more formidable barrier than the western face. The central part of the range, between Sacred and Harrison passes, shows the strongest and most rugged relief. South of Harrison Pass the crest of the range is generally even, though some high peaks break the level sky line.

In practically every canyon and gulch which heads in these mountains there are streams of good water, which is not the case in many of the ranges of Nevada. The streams usually sink a short distance after entering the valleys.

GEOLOGY.

AGE OF CERTAIN ROCKS.

The greater part of the east side of Ruby Range north of Harrison Pass is composed of light-colored granitic rocks that in many places exhibit a noticeable gneissic structure but in other places are porphyritic and show no indication of banding. The geologists of the Fortieth Parallel Survey regarded these granites, as well as a considerable thickness of hornblende and quartzitic schists that overlie

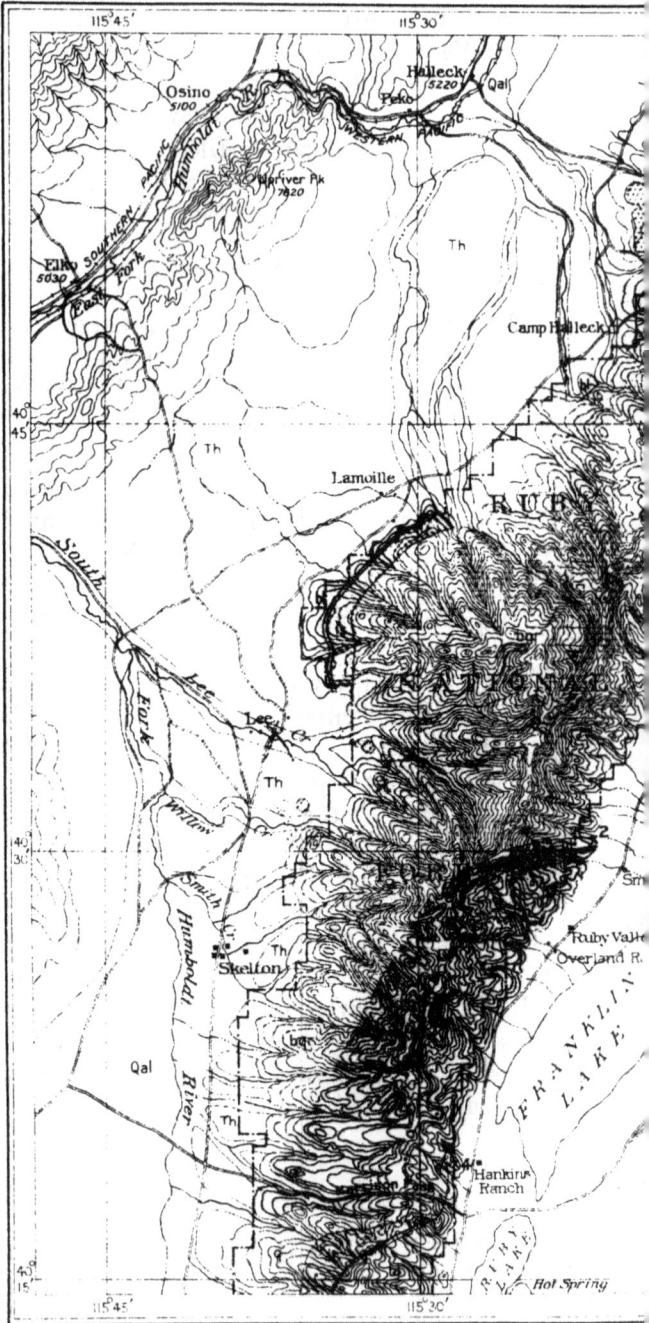

GEOLOGICAL MAP OF RUBY RANGE A
SOUTHWESTERN ELKO COUN

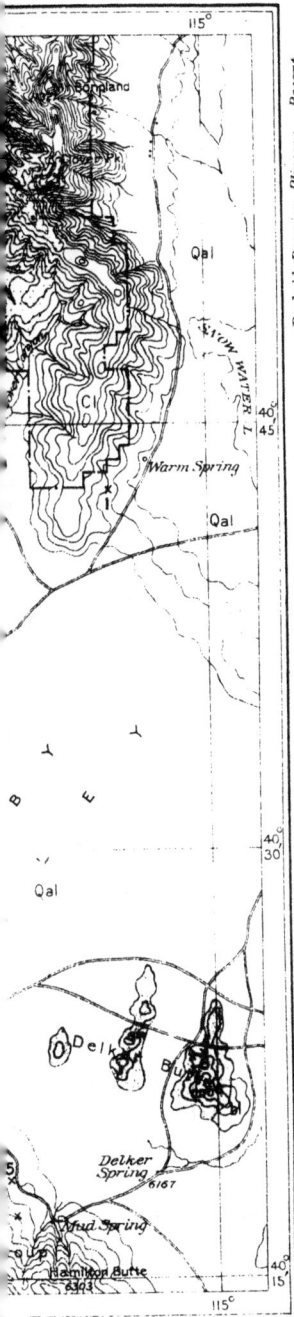

LEGEND

SEDIMENTARY ROCKS

Qal

Gravel and silt

Recent

Th

Humboldt formation
*(Partly consolidated
gravel and silt)*

Probably Pliocene

Cl

Buff and drab limestone,
some shale and quartzite

*Probably Permian
and Pennsylvanian*

bl

Blue limestone; some dolomitic
*Probably includes Carboniferous
(Mississippian), Devonian and Silurian,
and may include some Ordovician*

Owq

White quartzite
(Eureka quartzite?)

ls

Limestones and shales
(Limestones dark colored at top,
lighter colored below)

q

Quartzite
(Prospect Mountain quartzite?)

CHIEFLY IGNEOUS

Tr

Rhyolite

qm

Quartz monzonite
(Intrusive)

bgr

Biotite granite
*(Intrusive. On west side of range
includes quartzite and lime
schists which may be Cambrian)*

x²

Prospect
*(Number refers to following
list)*

1. Polar Star
2. Friday
3. Short
4. Valley View
5. Deadhorse
6. Delker

Right-side geologic age column:
QUATERNARY
TERTIARY
CARBON-IFEROUS
ORDOVICIAN? TO CARBONIFEROUS
ORDO-VICIAN?
PROBABLY CAMBRIAN AND ORDOVICIAN
PROBABLY CAMBRIAN
POST CARBONIFEROUS TERTIARY
*(May be Cretaceous
or early Tertiary)*

the granites on the western flank of the range, as of Archean age.[1] At a number of places on the flanks of the mountains isolated bodies of Paleozoic sediments rest upon the granite. A portion of sheet 4 of the atlas of the Fortieth Parallel Survey, including the Ruby Range, is shown in Plate III. The mapping of the geology on the eastern side of the mountains is somewhat modified as the result of the present reconnaissance.

SEDIMENTARY ROCKS.

On the eastern side of the Ruby Range (see Pl. III) there are three areas of sedimentary rocks. The largest of these is the outlying range of hills east of Clover Valley shown in the upper right corner of the map. On the atlas of the Fortieth Parallel Survey these hills are shown as Archean. They are composed of thin-bedded eastward-dipping limestones. Fossils collected at the Polar Star mine, at the southeast end of the ridge, indicate that the beds at that place are probably of Permian age. This collection was examined by G. H. Girty, of this Survey, who has identified the following species:

Batostomella sp.   Spiriferina pulchra?
Productus nevadensis?   Composita subtilita var.
Productus subhorridus?

It seems probable that some of the beds to the north and west of the Polar Star may be Pennsylvanian, though it is thought that all the beds on the east side of this ridge are Permian.

The small area of sediments northwest of Smith's ranch, in the central part of the range, is much metamorphosed, and its age was not definitely determined. At the south there is a belt of light-brownish quartzite dipping north, overlain by white crystalline limestones that are thought to be Ordovician. Both rocks have been intruded by fine-grained biotite granite which is somewhat gneissic.

West of Hankins ranch, opposite the south end of Franklin Lake, is the northern end of the body of Paleozoic sediments which form most of the Ruby Range south of Harrison Pass. The following description of this series is adapted from Hague's report.[2]

The sediments south of the area shown on Plate III dip east at low angles. The lowest beds exposed on the west side of the range are white to brownish-white vitreous quartzite. The transition from the quartzites to the limestones next above is made in a very short distance. The limestones are light gray to gray buff for a thickness of 1,600 feet, to a line where the color becomes darker

[1] Hague, Arnold, U. S. Geol. Expl. 40th Par. Rept., vol. 2, pp. 532–541, 1877.
[2] Idem, pp. 529–531.

and the bedding much more massive. The limestones of the lower portion are interbedded with arenaceous shales; the upper massive beds are free from shales, and some beds are dolomitic. A bed of white quartzite from 200 to 400 feet thick overlies the limestone and is followed above by the dark-colored massive limestones that form all the eastern part of the range south of Harrison Pass. The geologists of the Fortieth Parallel Survey reported " Lower Coal Measures " fossils from the later limestones.

Spurr[1] notes that the lower quartzites may be the equivalents of the Cambrian quartzites of the Snake Range in eastern Nevada. It is the writer's opinion, based, it is true, on very slender evidence, that the lower quartzite is of Cambrian age and probably to be correlated with the Prospect Mountain quartzite, and that the overlying limestone formations include the whole series from Cambrian to lower Carboniferous. He is also of the opinion that the quartzite and overlying dark limestone seen west of Hankins ranch should be correlated with the Ordovician Eureka quartzite and the Lone Mountain limestone of the Eureka section.

The series of dark limestones, which is succeeded below by lighter limestones and interbedded shales, he is of the opinion are of Cambrian and Ordovician age, and probably include representatives of the Eldorado limestone, Secret Canyon shale, Hamburg limestone, Dunderberg shale, and Pogonip limestone of the Eureka district.

The broad level of Ruby Valley east of the range is broken by two large swamp areas called Franklin and Ruby lakes. At some seasons of the year these swamps are under water, and the southern, Ruby Lake, usually contains water. These two depressions have the characteristics of playa lakes, which are common in the desert country, but they differ somewhat from most flats of this type in that they receive more water than most of the Nevada playas. The valley is filled with fine sands and silts, and much of the northern part of the valley could probably be " dry farmed " to advantage. The soil is rich, as is shown by the luxuriant growth of sage and greasewood, and many areas of rabbit brush bear testimony to the presence of water not far below the surface. Resting upon these silts along the base of the mountains there are small talus fans composed of bowlders of biotite granite mixed with finer sands and soil.

In the valley west of the range there are great deposits of partly consolidated, horizontally bedded sands and fine gravels of the Humboldt formation, which near the mountains grade into a coarser material. The bedding dips toward the axis of the valley at low angles.

[1] Spurr, J. E., Descriptive geology of Nevada south of the fortieth parallel and adjacent portions of California : U. S. Geol. Survey Bull. 208, p. 60, 1903.

IGNEOUS ROCKS.

In the Ruby Range two distinct types of igneous rocks are exposed, whose distribution is roughly indicated on Plate III. The granite underlying the east side of the northern part of the mountains is gneissic in most exposures, though in some places it does not exhibit banding. It is a very siliceous rock and in most of its facies quartz is the most abundant mineral. Orthoclase is the common feldspar, though some thin sections show microcline, and one contains a small amount of oligoclase. Brownish-green biotite is the common ferromagnesian mineral, though a little light-green hornblende is present in some specimens of the rock. Apatite and magnetite are the common accessory minerals, but magnetite is not abundant. In some varieties of the granite dark minerals are practically absent, but there are all gradations from these aplites through muscovite granite to biotite granite. Gneissic structure is common, due to the more or less parallel orientation of the mica flakes. There are, however, many exposures of fine to medium grained biotite granite and not a few bodies of pegmatitic granite.

These granites, which have been considered Archean, are intrusives, as is shown at the Friday and Short mines in Smith Creek, and at Sacred Pass. In these places dikes of the granite cut the sedimentary beds, and at Smith Creek belts of lime silicate minerals are developed along and near the contacts. In this locality, even in narrow granite dikes, there is a suggestion of gneissic structure. In his description of the Humboldt (Ruby) Range north of Fremont Pass, Hague [1] often alludes to garnet, fine black threads of hornblende, and prisms of actinolite developed in quartzites which form the western part of the range. His description casts considerable doubt upon his interpretation of the age of the granite. To be sure, its exact age is not known, yet it seems to be intrusive into early Paleozoic sediments. This rock does not resemble any of the late Cretaceous intrusives seen by the writer.

The "granite" of Harrison Pass, which extends northward to the vicinity of Smith Creek, was mapped by the geologists of the Fortieth Parallel Survey as an eruptive rock. It is a porphyritic quartz monzonite like that in many areas of the intrusive rocks of Cretaceous or Jurassic age in the State. Large phenocrysts of orthoclase are commonly well developed in a coarsely crystalline groundmass of feldspar, quartz, and biotite. Orthoclase and oligoclase-andesine are nearly equal in abundance in the rock, and together form at least two-thirds of its bulk. Flakes of dark-brown biotite are common, but some facies carry also hornblende.

---

[1] Hague, Arnold, U. S. Geol. Expl. 40th Par. Rept., vol. 2, pp. 532–537, 1877.

The relations between this porphyritic quartz monzonite and the biotite granite mass of the north end of the range is not known. The scant evidence gathered during this reconnaissance seems to indicate that one may grade into the other. Both varieties show a marked westward-dipping jointing, which parallels the trend of the range, and in the porphyritic quartz monzonite some masses exhibit a rough gneissic banding.

ORE DEPOSITS.

LOCATION.

So far as known there are very few prospects in the north end of Ruby Range. Three places on the east side of the range that show some mineralization were visited by the writer, and at a fourth, on the west side of the mountains, a little northwest of Smith Creek, there are said to be deposits similar to those seen at the Short mine.

HISTORY.

The earliest discovery of mineral on the east side of the Ruby Range was made at Cave Creek, about 8 miles south of Harrison Pass. According to White[1] the northward-trending veins in limestone carrying silver chloride and copper and lead carbonates were discovered in May, 1869, by Gen. Ewing, of the United States Army.

The next discoveries were the lead-zinc-copper ores of Smith Creek, which were opened by the Short brothers in 1903. The Short mine and the Friday properties, at the mouth of Smith Gulch, have produced a small amount of lead-silver ore. In September, 1913, the Short mine was under lease, and no doubt a small quantity of sorted lead-silver and zinc ore was shipped during the last year from this property.

TYPES OF DEPOSITS.

The deposits at the three localities visited are of distinctly different types. In the Warm Creek region lead and zinc carbonate ores are found in small, irregular replacement bodies along northwestward-trending fractures that cut limestones of probable Permian age. At Smith Creek the deposits of lead and zinc sulphides and copper carbonates and sulphides occur in replacement bodies associated with lime silicate minerals characteristic of zones of contact metamorphism. West of Hankins ranch scheelite and bismuth have recently been discovered in lenslike bodies of contact-metamorphic origin, which occur in early Paleozoic sediments intruded by quartz monzonite porphyry dikes.

---

[1] White, A. F., State Mineralogist of Nevada Third Biennial Rept., for 1869-70, p. 62 [1871].

Polar Star.—The Polar Star group of eight claims (No. 1, Pl. III) was located, according to the notice at the main work, on March 15, 1912, by T. J. Franks, M. E. Reed, and B. Woolverton. These claims are at the southeast side of the Warm Creek ridge, which lies east of Clover Valley. The main shaft is 5 miles southwest of Warm Creek ranch. The ridge is formed of light-gray, buff-weathering fossiliferous limestone and shales of probable Permian age. The beds strike N. 10° W. and dip 20° E. The principal development, on flat ground at the east base of the ridge, consists of an inclined whip shaft 64 feet deep, which has a few feet of water in the bottom. This shaft is sunk between two minor faults, which strike N. 45° E. and dip at 68° NW. The slips are 4 feet apart at the surface and are marked by slickensided faces and narrow clay seams. The fossiliferous limestone between these planes is only slightly brecciated, and in places large, white calcite crystals have been developed. At a depth of 50 feet there are drifts 40 feet northeast and 20 feet southwest along the zone. In the face of the southeast drift the hanging-wall slip dips to the northwest at 40°, and the footwall slip straightens to a dip of 85°. At the top of the face, between the two fractures, there is a small opening in which a little lead carbonate and zinc carbonate has been deposited. At several places along the hanging-wall slip in other parts of the work there are small kidneys and stringers of lead and zinc carbonate ores, which occur as replacements in the slightly brecciated limestone.

Other prospects.—Several pits and shafts have been sunk on what appears to be the same fractures southwest of the main workings. A little work has been done also on a small body of replacement carbonate ore west of the summit of the low ridge about half a mile west-northwest of the Polar Star incline.

Location and features of the deposits.—The Ruby Valley district (No. 9, fig. 1), so far as could be learned, includes the mines in Smith Creek, whose mouth is about 2 miles northwest of Smith ranch and 6 miles north of Ruby Valley post office. (See Pl. III.) The principal properties are the Friday group of nine claims, on the north side of the mouth of the canyon, and the Short group, at the forks of Smith Gulch, about 2 miles west of Ruby Valley. The Friday group was located in 1906 and was acquired by the Friday Gold Mining Co. soon afterward. The Short group of seven claims was located by J. F. and A. M. Short in 1903. The Michigan claim of this group is at present bonded to the Arthur Zinc Mining Co.

The mines of this district occur in a narrow belt of white crystalline limestones which strikes about N. 60° E. and dips 45° to 60° N. The bedding has not been entirely obliterated and seems to be parallel to the strike of the belt. The limestones continue to the summit of the range, and are said to extend across the mountains. Some lead prospects 12 miles southeast of Lee post office, on the west side of Ruby Range, are said to be in the same belt. These limestones have been intruded by a fine-grained biotite granite, consisting essentially of quartz, orthoclase, dark greenish-brown biotite, and a minor amount of oligoclase. The intrusion of this granite has metamorphosed the limestone at a number of places into lenses of lime silicate rock consisting largely of diopside and tremolite, with a little quartz. Phlogopite, the brown magnesium-bearing mica, was determined microscopically in specimens from one locality. Garnet does not seem to have been developed to any considerable extent. Some of these lenses contain galena, sphalerite, and more rarely chalcopyrite. So far as developed the ore bodies appear to be neither large nor persistent in any direction.

Friday group.—The main development on the Friday group consists of a tunnel and open cut at the northeast base of the ridge north of Smith Gulch. The tunnel runs S. 70° W. for about 400 feet through a marmarized limestone, cut in an intricate manner by a fine-grained light-gray granite. The granite consists essentially of quartz, orthoclase, and brownish-green biotite, the quartz constituting over two-thirds of the rock mass. The limestone in the ridge is a coarsely crystalline white rock, which near the granite has been metamorphosed to a brownish-pink aggregate of calcite and phlogopite and some quartz. In the small lenslike ore bodies greenish tremolite in large, pearly radial forms is intergrown with quartz and the sulphides. Galena is the most abundant sulphide, though copper sulphide must have been present, as there is a small amount of copper carbonate in the surface ore. Cerusite and anglesite coat some of the galena.

The Crescent workings, in flat ground, about one-fourth mile east of the tunnel, consist of a shallow shaft on a vertical fracture trending N. 15° W. and cutting white crystalline limestone and lime silicate rock. No igneous rocks are exposed in the immediate vicinity, yet the limestone is altered to a heavy green rock consisting almost wholly of diopside and tremolite. Sulphides are not abundant here, but a small amount of copper carbonates occurs along the fissure, and specks of bornite surrounded by iron oxide and malachite are seen in the rock near the fracture.

Short group.—The Short group covers a limestone spur between the two forks of Smith Gulch. From the north fork a tunnel 250 feet long has been driven S. 85° W. through white crystalline limestones, which apparently dip north-northeast at fairly steep angles. Three dikes of biotite granite are exposed in the tunnel. The lime-stones are only slightly metamorphosed, even at the sharp contacts, though some brown garnet was noted along one dike in which the granite for an inch from the contact contains practically no mica. On the surface south of the tunnel mouth large masses of tremolite and diopside occur along what appear to be the same dikes that are exposed in the tunnel. Galena, sphalerite, and a little chalcopyrite, with their oxidation products, are sparingly distributed through all these lenses of lime silicate rock.

The Michigan incline is 80 feet deep and dips 45° N. A 50-foot drift has been driven to the west from its bottom. This incline is about parallel to the bedding of white crystalline limestone, and is only a few feet north of a contact of limestone and granite. The contact is intrusive, and in the workings there are a number of small dikes. The ore makes in lenslike masses parallel to the bedding, which consists of radial tremolite, crystalline galena, and dark-brown sphalerite. Some bodies are largely sphalerite; others are nearly pure galena; yet both minerals seem to be present in all the ore. The galena ore is said to carry about 4.5 ounces of silver per ton. Along fissures cutting both wall rock and ore there has been a slight amount of oxidation, but sulphides surrounded by thin crusts of anglesite and cerusite were found directly at the surface.

In a near-by prospect chalcopyrite as well as galena and sphalerite is present, though the copper minerals do not seem to be particularly abundant in this vicinity.

VALLEY VIEW DISTRICT.

Hankins property.—The Valley View prospect (No. 12, fig. 1, p. 18), known as the Hankins property, after its discoverer, is on the eastern side of the Ruby Range, 10 miles south-southwest of Ruby Valley post office, opposite the south end of Franklin Lake. (See Pl. III.) Native bismuth was discovered here early in 1913, and the de-velopment on the group of 10 claims consisted at the time of visit of a few shallow shafts and a number of trenches. Atkins, Kroll & Co., of San Francisco, Cal., who had the property under bond in September, were prospecting for scheelite with fair results.

The deposit occurs in white crystalline limestone, a belt of which about 600 feet wide lies between quartzite to the west and north and dark contorted arenaceous limestone to the south. A number of small faults cut these formations, and locally the dark, thin-bedded limestones show intricate contorted folds. The edge of the main mass of intrusive porphyritic quartz monzonite is nearly half a mile west of the prospect. Offshoots from this intrusive have invaded the sedimentary rocks, usually as narrow dikes parallel to the major structure. Most of the dikes are fine to medium grained porphyritic quartz monzonite containing few ferromagnesian minerals. Quartz, orthoclase, and greenish-brown biotite are the phenocrystic minerals and, together with oligoclase, make up the groundmass. One dike apparently grades into a dark porphyry with prominent altered plagioclase phenocrysts set in a groundmass of labradorite, green amphibole, biotite, and quartz.

Along these small dikes, which ranged in width from a few inches to 8 feet, there has been as a rule only slight contact metamorphism. In the dark arenaceous limestone chlorite, epidote, and little light-green pyroxene have been developed for a few inches from the contacts. The white crystalline limestone has been more intensely altered, though the lime silicate zones are narrow and are not particularly noticeable except on close inspection. Some dikelike bodies consisting of phlogopite, rutile, and light-green pyroxene seem to represent highly metamorphosed beds of limestone rather than intrusive bodies.

Small lenslike bodies of ore are exposed in a number of pits. Most of these bodies are near dikes and, like them, parallel the general structure. They usually consist of two more or less well-defined zones. The outer, which is frozen to the coarse calcite crystals of the walls, consists of an intergrowth of pod-shaped white crystals of scheelite, the largest half an inch in diameter, of large flat greenish-yellow crystals of epidote, and of quartz, chlorite, and calcite. Thin sections of this material studied under the microscope show that the scheelite at least is a metasomatic replacement of calcite. Near the center of the lens native bismuth and pyrite begin to appear. The central part of the lenses are usually fine-grained aggregates of quartz, phlogopite, and light-green to white pyroxene, in which there are small crystals of pyrite, bismuthinite (Bi_2S_3), and native bismuth. Whether these aggregates were originally dikes is not certain, though in one pit the suggestion was strong that an 8-inch highly altered dike formed the center of the vein. In another pit the center of the lens consisted of an aggregate of quartz, phlogopite, and pyroxene, with the metallic minerals sparingly distributed through it.

The development was not sufficient at any place to warrant a statement as to the probable continuance of these deposits with increase of depth, but the fact that any particular lens of ore does not extend for many feet on the surface would raise the question whether any large body of ore would be found.

MUD SPRINGS DISTRICT.

LOCATION AND ACCESSIBILITY.

The Mud Springs district (No. 8, fig. 1, p. 18), sometimes called the Medicine Springs district, is at the north end of the Ruby Hills, which form the southeast side of Ruby Valley. The prospects are in the low hills west of Mud Springs Canyon and 6 miles northwest of Hamilton Butte, the highest point of the range. They are about 20 miles by road southeast of Ruby Valley post office. Currie, on the Nevada Northern Railway, the nearest shipping point, is about 40 miles away by road toward the east-southeast. Wells, on the Southern Pacific, is 60 miles north of the main camp. There is a good supply of water at Mud Spring, 2 miles to the east, and at Medicine Spring, 5 miles northwest of the Deadhorse camp.

GEOLOGY.

All the north end of the Ruby Hills is composed of drab and cream-colored limestones, cherty in many places, in beds 2 to 4 feet thick. In the vicinity of the Deadhorse camp a 75-foot bed of yellow calcareous shale and a 25-foot bed of light buff-colored sandy quartzite are interbedded with the limestone. In all the low hills north and west of Hamilton Butte the limestones dip 10°–15° E. Hague[1] reports that in Hamilton Butte there is a very low westward dip, showing a general shallow synclinal structure for the whole group. He noted the presence of fossils in many of the limestones and says that "they probably represent the upper horizon of the Wahsatch limestone." The collections of fossils made by the writer at Medicine Spring, at a prospect a mile south-southwest of Deadhorse, and a short distance northwest of the Deadhorse shaft, have been examined by G. H. Girty, who reports the following species:

Derbya? sp.	Chonetes utahensis.
Batostomella sp.	Productus (Aulosteges?) sp.
Productus nevadensis?	Spirifer sp.
Lioclema sp.	Spiriferina pulchra.

These fossils, he says, denote a Permian age for the beds. It seems probable that all the formations in the northern part of these hills are of the same age, yet some of the limestones along the western side of the group may be upper Pennsylvanian.

ORE DEPOSITS.

History and development.—It is said that the first discoveries of mineral in this vicinity were made in 1910 by Sam Backman, Garfield Bardness, and Fred Martin. The Deadhorse claim, which has

---

[1] Hague, Arnold, U. S. Geol. Expl. 40th Par. Rept., vol. 2, pp. 490–491, 1877.

more development than any other, was located in January, 1911. It is now owned by the Nevada Dividend Mining Co., which has sunk a vertical shaft 114 feet deep. A station at the 100-foot level had been cut and sinking was in progress at the time of visit (September, 1913). Hoisting is done by bucket and whim. On most of the other claims in the district little more than location work had been done. No ore had been shipped from the camp up to September, 1913.

Character of the deposits.—The deposits are replacements of limestone along small brecciated fractures which strike north to N. 35° E. and dip east at steep angles. In most of the open cuts limonite and more or less lead and zinc carbonates and barite constitute the ore. At the Deadhorse property residual galena has been found irregularly scattered through the ore, which is largely oxidized to a depth of 114 feet. At a number of places there are fair surface indications, but it seems doubtful if the metallization is extensive. The hills have a relatively undisturbed geologic appearance.

Deadhorse property.—The Deadhorse group of five claims (No. 5, Pl. III) is in a saddle southeast of a small butte which is capped by brownish-colored sandy quartzite. Under the quartzite there are 75 feet of yellow to buff calcareous shales, which rest upon the buff and drab fossiliferous limestones in which the ore occurs.

The ore follows a N. 35° E. fracture which dips 80° E. The croppings are 2 to 3 feet wide, stand 1 to 2 feet above the surface, and are traceable for over 1,500 feet along the strike of the ledge. They consist of calcite, quartz, and barite, with irregularly distributed small pockets of lead carbonates and sulphates.

At the 100-foot level a subparallel fracture 12 feet from the main one is exposed in an east drift. The limestone between the two is somewhat brecciated, but no marked movement is apparent along either slip. In each the brecciated limestone next to the walls is slightly iron stained. On the hanging-wall side there is a foot of ledge matter carrying small irregular pockets and stringers of galena surrounded by anglesite and cerusite, and on the footwall side there is 4 to 6 inches of iron-stained material. Barite is particularly abundant in the vicinity of the small lead pockets.

Other prospects.—An open cut and shallow shaft on the Madison claim (relocated Jan. 1, 1912, as the Jackpot) is 1 mile south-southwest of the Deadhorse. The light brownish-gray fossiliferous limestones dip 15° E. in this locality. A fracture with a north-south strike and a dip of 80° E. cuts the beds but shows little or no movement. For 2 feet on either side of the open crack the limestone is altered to a porous mass of iron oxide carrying some cerusite and smithsonite, and in some of the cavities small well-developed crystals

of calamine and cerusite. Barite is present in varying quantities.
No sulphides are exposed in the shallow cut.

South and east from the Deadhorse shaft a number of shallow
pits and open cuts disclose oxidized lead-zinc ores, which occur as
replacement deposits along more or less vertical, north-striking
fissures.

DELKER DISTRICT.

LOCATION.

The Delker district (No. 2, fig. 1, p. 18) is on the northeast side of
Delker Hill, the easternmost of the low hills which rise in the plain
east of Ruby Lake and which were called the Franklin Buttes by the
geologists of the Fortieth Parallel Survey. These hills are in latitude
40° 25' N. and extend from longitude 115° to 115° 7' W. (See Pl.
III.) Currie, on the Nevada Northern Railway, the shipping point,
is about 25 miles southwest of the mines.

GEOLOGY.

According to Hague[1] the buttes are for the most part composed
of granitoid rocks. Hague notes, however, the presence of "Lower
Coal Measure" limestones at the southeast side of the eastern butte.
He says: "* * * The rocks which form these buttes present an
interesting gradation from a syenitic granite, through granite porphyry, into genuine felsite porphyry." From the description of these
various phases of intrusive rocks, all of which contain both orthoclase and plagioclase feldspar with quartz and biotite, it would seem
that they are best referred to the quartz monzonite series, which at
many exposures in the State shows the gradations noted by Hague.
A thin section cut from a specimen of the intrusive rock collected
by the writer near the prospects at the northwest side of the east
butte, shows an inequigranular rock consisting essentially of feldspar
with quartz, brown biotite, and a little light-green hornblende.
Oligoclase-andesine exceeds the orthoclase in amount in this particular specimen. It is not porphyritic, though farther south along
the west side of the butte some rock which appeared to carry the
same minerals was porphyritic.

ORE DEPOSITS.

History and economic conditions.—So far as could be learned
copper ores were first discovered at Delker in 1894. A little prospecting has been done annually since the discovery, but it is said
that no ore has been shipped. It is understood that the deposits are

---

[1] Emmons, S. F., U. S. Geol. Expl. 40th Par. Rept., vol. 2, pp. 491–493, 1877.

largely controlled by J. P. Stratton, of Cherry Creek, and William Griswald, of Ruby Valley.

Water has to be hauled from Delker Spring (shown on sheet 4 of the atlas of the Fortieth Parallel Survey as Locust Spring), at the south end of the east butte, or from a spring on M. L. Atwood's ranch, about 5 miles east of the mines.

A little juniper grows on the buttes, but timbers for mining use must be hauled from Spruce Mountain, 25 miles north-northeast.

Types of deposits.—At the prospects the quartz monzonite is intrusive into some poorly exposed limestones and interbedded quartzites. Lime silicate zones are developed at a number of places along the contact, and copper carbonate ores were seen in a number of prospect pits associated with the contact-metamorphic minerals. One of the larger workings on the contact is a 50-foot incline shaft that follows the quartz monzonite footwall, which strikes N. 70° W. and dips 60° N. The intrusive rock also carries a few veinlike deposits along fractures that strike approximately N. 35° E. and dip 80° NW.

The chief metallic minerals seen are chrysocolla, malachite, and a small amount of azurite. Copper pitch ore is present. Some fairly large masses of greenish jasperoidal rock are probably mostly chrysocolla. In one specimen a very minor amount of the chalcocite was noted.

SPRUCE MOUNTAIN DISTRICT.

LOCATION AND ACCESSIBILITY.

The Spruce Mountain district (No. 10, fig. 1, p. 18) covers Spruce Mountain, the southwestern peak of the Gosiute Range, and lies a few miles west of longitude 114° 45' and north of latitude 40° 30'. The mines are on the west and north flanks of the peak, which, according to the geologists of the Fortieth Parallel Survey,[1] has an elevation of 10,400 feet above sea level. The principal supply points are Wells, on the Southern Pacific, about 45 miles north, and Currie, on the Nevada Northern Railway, 24 miles southeast. Freight is hauled to and from Tobar siding, on the Western Pacific Railway, 21 miles north.

TOPOGRAPHY.

Spruce Mountain is a somewhat isolated peak west of the axis of the main range, with which it is connected by comparatively low hills. It rises nearly 4,000 feet above the flat valleys which surround it on all but the eastern side, presenting a rugged topography, particularly to the south and east. Northward a long ridge extends for some miles into Antelope Valley.

---

[1] Hague, Arnold, U. S. Geol. Expl. 40th Par. Rept., vol. 2, p. 505, 1877.

GEOLOGY.

SEDIMENTARY ROCKS.

The larger part of the mountain is composed of dark blue-gray limestones in 1 to 4 foot beds, interbedded with some shales and quartzites. The structure is complicated by a number of faults, but northwest of Spruce Mountain, where the beds are least disturbed, it is estimated that the limestones are 2,000 to 3,000 feet thick. They were considered to belong to the "Lower Coal Measures" by the geologists of the Fortieth Parallel Survey,[1] who collected the following fossils on the summit and northwest side of the peak:

Productus costatus.	Eumetria punctulifera.
Productus semireticulatus.	Fusulina cylindrica.
Productus nebrascensis.	Trematopora.

Among a few fossils collected by the writer at the west base of the mountain and in the saddle between Spruce Mountain and Banner Hill (see fig. 4) Dr. Girty identified the following species, which he considers indicative of Mississippian age:

Crinoid fragments.	Amboccelia? sp.
Stenopora? sp.	Lithostrotion? sp.
Echinocrinus sp.	Chonetes platynotus?
Syringopora? sp.	

IGNEOUS ROCKS.

Near the summit of Spruce Mountain there are a number of small, poorly exposed bodies of igneous rock. The geologists of the Fortieth Parallel Survey[2] noted the presence of both basic and acidic intrusives.

On the north slope of Spruce Mountain, east of the saddle between that peak and Banner Hill, there is a small stock of porphyritic granite with prominent quartz and altered orthoclase phenocrysts. It is light in color, breaks in angular fragments, and contains few iron-bearing minerals, though it is somewhat iron stained on the surface. Thin sections of this rock examined microscopically show the groundmass to be composed of a granular intergrowth of quartz and a feldspar that seems to have been largely orthoclase but that is now entirely altered to sericite, quartz, and calcite. No original ferromagnesian minerals are seen in the thin section, though a little chlorite and iron oxide are present. There are other small dikes of similar rock, though no body so large as the one just described was noted elsewhere in the district.

---

[1] Hague, Arnold, op. cit., pp. 510-511.   [2] Idem, pp. 511-512.

On Banner Hill and the ridge west of it there are a few small dikes of diorite and diorite porphyry. Wherever seen this type of igneous rock is highly altered. The most altered rocks are brownish,

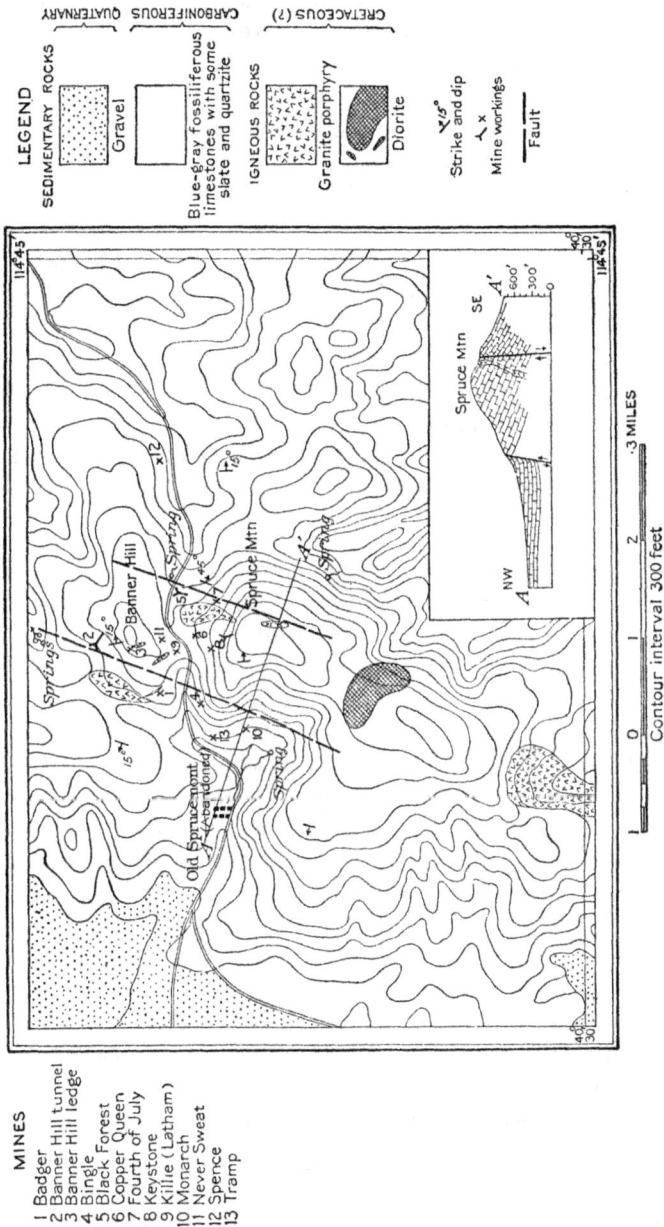

LEGEND

SEDIMENTARY ROCKS

QUATERNARY — Gravel

CARBONIFEROUS — Blue-gray fossiliferous limestones with some slate and quartzite

IGNEOUS ROCKS

CRETACEOUS (?) — Granite porphyry — Diorite

Strike and dip

Mine workings

Fault

Spruce Mtn

Contour interval 300 feet

MINES
1 Badger
2 Banner Hill tunnel
3 Banner Hill ledge
4 Bingle
5 Black Forest
6 Copper Queen
7 Fourth of July
8 Keystone
9 Killie (Latham)
10 Monarch
11 Never Sweat
12 Spence
13 Tramp

FIGURE 4.—Sketch map of the Spruce Mountain district, Elko County, Nev., showing positions of principal mines and indicating the geology of the region.

rather soft, and easily eroded; less highly altered holocrystalline rocks are greenish brown and of medium grain. The hornblendes are readily determined in the field, though they show some alterations, but the feldspars are largely altered. Thin sections of the rocks from two different bodies show well-crystallized, dark greenish-brown hornblende altering to chlorite and epidote, intergrown with feldspar, which in both slides is so much altered that its character can not be determined. Its outlines, however, suggest plagioclase. The alteration products are calcite, sericite, and iron oxide. A small amount of quartz is present in both rocks, but most of it is believed to be secondary. Apatite is abundant, particularly in the dike northwest of the Killie shaft.

CONTACT METAMORPHISM.

The limestones have not suffered much contact metamorphism along these dikes, though there is a 2-foot zone of greenish lime silicate rock along a dike on Banner Hill, and in the long drift from the Killie shaft there is a 12-foot zone of mineralized silicate rock between the granite porphyry and limestone. Light-greenish garnet is the most common metamorphic mineral, though a little diopside is present in the contact zone near the diorite dike on Banner Hill.

STRUCTURE.

Spruce Mountain seems to have been formed as an anticlinal fold but to have been much modified, in the vicinity of the mines, by the intrusion of the igneous rocks and by two nearly parallel normal faults, striking approximately N. 20° E. and dipping as shown in figure 4. The western fault seems to have been the larger, and is marked by a rather prominent scarp and by several springs along its strike. A line of secondary faulting is seen on the west slope of Banner Hill. The eastern major fault is shown in a saddle on the east side of the summit of Spruce Mountain, and what is probably the same fault is cut by workings at the Black Forest mine. Both of these faults are thought to have steep dips. (See fig. 4.)

The position of the granite porphyry intrusives is near the fault zones. Underground, in the Killie shaft, the porphyry mass, apparently the same as that near the Black Forest and Fourth of July mines, extends farther west and north than it does on the surface.

West of the west fault, between Old Sprucemont (abandoned) and the divide in which the mines are located, the limestones lie essentially horizontal, though showing some minor breaks. Between the two major faults the beds, owing to minor faults and the intrusion of the igneous rocks, show considerable diversity in dip, inclining northeast, east, or southeast at low angles. Eastward from the east fault the limestones dip east, at first steeply but within a short distance only moderately.

A fault which strikes N. 35°–40° E. and dips 50° NW., not exposed on the surface, is seen in the Keystone and Copper Queen workings. It is a strong normal fault with a belt of crushed limestone 30 feet wide along its northwest side. The hill northwest of the fault is covered with a deep accumulation of talus, soil, and rubble, as is clearly shown in the Copper Queen tunnel and in the northern surface workings of the Keystone.

MINERAL DEPOSITS.

HISTORY AND PRODUCTION.

According to Raymond[1] the discovery of lead-silver ores at Spruce Mountain was made in 1869 at the Latham mine. In 1871[2] the three original districts—Latham, Johnson, and Steptoe—were combined to form what is now the Spruce Mountain district, which includes all of the mines in this vicinity. The principal mines in the early days were the Latham, Juniper, Fourth of July, and Black Forest. The Ingot Mining Co. acquired the Latham mine and several others in 1871, and in 1872 built a 4½-foot Philadelphia type smelter at Sprucemont to treat the lead carbonate ores. This furnace never proved a success, partly because of lack of iron for fluxing and small supply of water and partly because of the presence of considerable copper and not much silver in the ore. For a short while 35 tons of ore were smelted daily. The furnace was shut down in 1873 and, so far as could be learned, has not been "blown in" since. At present there is a 50-ton furnace belonging to C. M. Spence in the lower part of the valley northeast of Spruce Mountain. It is said to have smelted a considerable tonnage of ores from the Black Forest mine for some years prior to 1910.

Just previous to the panic of 1907 it is said that there was a short-lived revival of activity in the camp. In September, 1913, less than 10 men were operating in the district and only one small shipment had been made by lessees on the Keystone property during the year.

There are no reliable figures of the early production from the various mines. One estimate of the total output from the Monarch, Latham, Killie, Banner Hill, Black Forest, Spruce, Juniper, and Fourth of July is $700,000. The following table of production for the period from 1902 to 1912, inclusive, is taken from the Mineral Resources reports of the United States Geological Survey.

---

[1] Raymond, R. W., Statistics of mines and mining in the States and Territories west of the Rocky Mountains for 1870, p. 152, 1872.

[2] Idem for 1872, pp. 160–161, 1873.

Production of the Spruce Mountain district, Elko County, Nev., 1902–1912.

Year.	Gold.	Silver.	Copper.	Lead.	Total value.
	Fine ounces.	*Fine ounces.*	*Pounds.*	*Pounds.*	
1902		189,072		1,272,600	$145,440
1903		29,175	6,000	451,000	27,063
1904	62.73	8,773		122,869	9,954
1905	1.16	10,722		185,257	16,177
1906		3,648		59,965	5,862
1907	3.43	1,806	116,592		24,590
1908		75	10,977		1,489
1909	5.56	10,668	40,615	201,861	19,622
1910					
1911					
1912		14		13,277	605
	72.88	253,953	174,184	2,306,829	250,802

OCCURRENCE OF THE DEPOSITS.

The ore deposits are, with a few minor exceptions, replacements in limestone, closely associated with fractures and faults and with the intrusive igneous rocks. At a few places copper ores occur with lime silicate minerals of contact-metamorphic origin. The replacement deposits are of two types, bedded and fissure. Most of the production of the district has come from bedded deposits, but the fissure replacement deposits are possibly of more importance at present.

THE ORES.

Only carbonate ores have been mined up to the present time, and lead carbonate ores have been more profitable than those carrying copper.

The copper-bearing bodies are found north of the saddle between Spruce Mountain and Banner Hill and seem to be smaller than the lead bodies south of the saddle. The minerals of the copper ores are malachite, chrysocolla, and chalcopyrite. Minor amounts of bornite and chalcocite were noted in ores from Banner Hill, and copper pitch ore is rather abundant at the surface. The ores from the old Latham mine contained considerable copper carbonate but were valuable for their argentiferous cerusite.

The lead-silver ores, all of which are limonitic, contain cerusite, anglesite, an oxidized antimony mineral, and residual kernels of galena. They seem to be more closely associated with the major faults and with the granite porphyry intrusives than with the copper-bearing ores.

THE PROPERTIES.

Badger tunnel (No. 1, fig. 4).—The Badger tunnel, 100 feet long, runs N. 16° W. along the contact of flat-lying limestone and lime shale with a diorite dike. It is about one-half mile west-northwest of the Killie shaft. At one place, near the face of the tunnel, a small lens of copper carbonate ore, apparently a bedded replacement, is said to carry lead and silver.

Banner Hill claims (No. 2, fig. 4).—A group of 23 claims controlled by the Johnson brothers, of Wells, covers the western part of Banner Hill, north of the Never Sweat property. There are a number of shallow surface workings at various places on the north and west slopes of the hill on deposits of copper carbonate ores, some of which are associated with small lenses of garnet and diopside, though the majority are bedded or fissure replacements. Remnants of chalcopyrite are commonly found even in surface ores, and bornite and chalcocite are occasionally seen as thin coatings between the chalcopyrite and carbonate ores. This group of claims is being developed through a crosscut tunnel, which starts in the head of Latham Canyon and runs S. 55° E., cutting limestones interbedded with dark quartzites. In September, 1913, it was 512 feet long. About 100 feet from the mouth an 8-foot dike of granite porphyry was cut; 450 feet from the mouth a very highly altered diorite dike is seen; and 200 feet from the mouth a minor north-south fault dips 50° W.

The Banner Hill ledge (No. 3, fig. 4), near the summit of the hill, is a zone of crushed, slightly garnetized limestone along a fracture that strikes N. 20° E. and dips 80° W. A lenslike body of copper carbonate ore 5 to 15 feet wide in this ledge is developed by a 100-foot drift tunnel and a 20-foot winze.

Bingle claim (No. 4, fig. 4).—The Bingle claim, three-fourths mile southwest of the Killie shaft, covers a siliceous iron-stained cropping 20 by 160 feet in area, whose longer dimension strikes N. 50° E. It is in limestones which are nearly horizontal and is probably a replacement body along a fissure.

Black Forest claim (No. 5, fig. 4).—The Black Forest claim is on the north slope of Spruce Mountain near the eastern fault and northeast of the intrusive granite porphyry. Development consists of four connected tunnels, of which only the lowest could be entered. This lowest tunnel runs 500 feet south of and parallel to the main fault, to which there are a number of crosscuts, and near which the ore seems to make. Three fractures trending N. 40°–50° E., exposed in the tunnel, are cut off by the main fault. The ore bodies were apparently located at the junctions of the cross faults with the main fault, and the ore seems to have replaced the crushed, sericitized, pyrite-impregnated limestone below the 6-foot gouge which marks the main fault. None of the stopes could be entered and but little ore remains on the dump. The ore seen, reported to be like most of that taken from the mine, which carried 20 per cent lead and 20 ounces silver a ton, is a fine-grained, soft ocherous material containing iron, lead, antimony, and silver combined as oxides, carbonates, and sulphates.

The individual grains can not be separated for definite determination, but under the microscope at least three distinct substances can
be seen. The most abundant consists of yellowish-white, minute
hexagonal crystals, the next of reddish-yellow indeterminable earthy
masses, and the third of some yellowish-green material.

Copper Queen mine (*No. 6, fig. 4*).—The Copper Queen workings
are between the Killie shaft and the Keystone, near the granite
porphyry stock. A vertical shaft 100 feet deep, with short levels
at 60 and 100 feet, and a 39-foot inclined winze from the east 100-
foot drift, which connects with a crosscut tunnel about 400 feet long,
constitute the workings. The relations are rather peculiar and are
probably best brought out in figure 5. At the southwest a well-

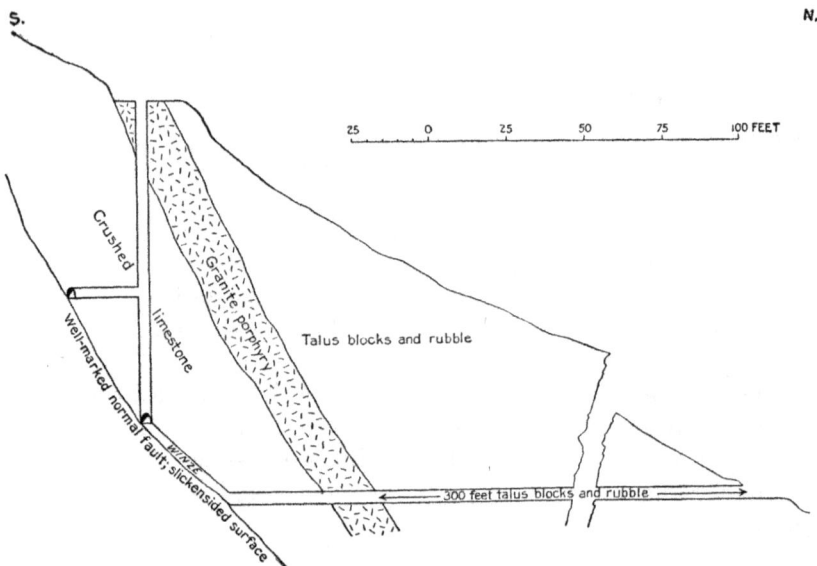

FIGURE 5.—Generalized elevation of the Copper Queen workings, Spruce Mountain district, Elko County, Nev.

marked normal fault plane that strikes N. 35° E. shows deep
grooves on the polished face parallel to the dip. Above this plane
is 12 to 18 inches of gouge followed by 25 to 40 feet of crushed limestone, which irregularly carries lenses of copper and lead carbonate
ores. Above, or northwest of the crushed zone, there is a 30-foot
dike of granite porphyry, which is seen at the collar of the shaft and
in the crosscut tunnel. The igneous rock is crushed to some extent
and contains a little pyrite. Above the porphyry the tunnel runs
through 300 feet of talus, dirt, and rubble such as would accumulate
at the base of a steep slope from ordinary weathering.

Fourth of July mine (*No. 7, fig. 4*).—The Fourth of July mine lies
just west of the eastern fault near the southern end of the granite por-

phyry stock on the north slope of Spruce Mountain. Whitehill[1] reported a 187-foot shaft and 694 feet of drifts on the property in 1873. In September, 1913, most of the old workings were caved, though a heavily timbered tunnel could be entered for 300 feet. None of the rock seems to be in place, but this effect might be due to intense shattering of the limestone and quartzite between the fault and the igneous stock. The silver-bearing lead carbonate ore is said to have occurred in small irregular pockets.

Keystone mine (No. 8, fig. 4).—The Keystone, on the north slope of Spruce Mountain about 300 feet above the pass, is worked through open cuts, shallow shafts, and short tunnels, largely in loose talus, though at places in the upper workings in limestone. At that shaft a fracture zone having an irregular steep dip either north or south strikes N. 40° E. In this fracture iron-stained cellular oxidized lead ores are found. Limonite, anglesite, cerusite, and residual kernels of galena were noted. From the talus below this fracture a considerable quantity of silver-bearing oxidized lead ore has been mined.

Killie (Latham) mine (No. 9, fig. 4).—The old Latham mine in the saddle between Spruce Mountain and Banner Hill was worked by tunnels and shallow shafts west of the Killie shaft. The old workings indicate that the ores occurred in bedded replacements of limestones and shale, which have a very low northeast dip. Raymond[2] reports that in 1872 the shaft was 108 feet deep with considerable drifting in the bedded deposit. The ore was silver-bearing lead carbonate but contained some antimony and copper. It is said to have averaged $50 to $60 a ton in lead and silver.

The Killie shaft, sunk by the Spruce Mountain Copper Co., is east of the old workings and at a slightly lower elevation. It was not entered but is reported to be 220 feet deep. At the 200-foot level it is said that a drift runs 1,450 feet southeast, toward the Keystone and Fourth of July. The first 1,100 feet of the drift is in limestones and the face in granite porphyry. Both rocks were seen on the dump, as was also considerable crushed greenish pyrite-impregnated limestone, which is said to occur, together with a 12-foot belt of pyrite-galena-chalcopyrite ore, at the contact of limestone and granite. It is said that for several feet northwest of the contact the limestones are more or less mineralized.

The shaft is equipped with a 25-horsepower gasoline hoist and compressor. A sawmill and blower are run by a separate engine covered by the shaft house.

---

[1] Whitehill, H. R., State Mineralogist of Nevada Fourth Biennial Rept. for 1871–72, pp. 24–28 [1874].

[2] Raymond, R. W., Statistics of mines and mining in the States and Territories west of the Rocky Mountains for 1872, p. 160, 1873; idem for 1873, p. 217, 1874.

Monarch claim (No. 10, fig. 4).—The old abandoned caved workings of the Monarch and Hartley claims are in the foothills about three-fourths of a mile east of Old Sprucemont. Large dumps at both places indicate rather extensive workings in what appear to have been bedded deposits occurring in nearly horizontal limestones near fractures. Little ore remains about the dumps; that seen is a limonitic lead carbonate.

Never Sweat prospect (No. 11, fig. 4).—The Never Sweat workings were not accessible. They consist of three shafts on a ledge striking N. 20° E. and dipping 65°–70° W. and cutting limestones interbedded with quartzite. The ore on the dump shows only copper carbonate minerals replacing and coating fragments of brecciated limestone.

Spence claims (No. 12, fig. 4).—Mr. C. M. Spence owns a large group of claims east of Banner Hill, which he is developing through a long drift from the 200-foot level of a vertical shaft about 2 miles east of the summit between Spruce Mountain and Banner Hill. The property was not visited, but it is reported that all of the work is along a contact of granite porphyry and limestone.

Tramp shaft (No. 13, fig. 4).—The Tramp shaft, about 1¼ miles east-northeast of old Sprucemont, is in the low flat hills, underlain by horizontal limestones and shales. It is 60 feet deep and is sunk on the south side of a fault trending N. 40° E., along which there has been much crushing of the sediments. Pyrite is abundantly disseminated in the breccia and is occasionally seen with chalcopyrite in small lenses and stringers cutting the crushed limestone.

DOLLY VARDEN (MIZPAH) DISTRICT.

LOCATION AND ACCESSIBILITY.

The Dolly Varden district (No. 3, fig. 1, p. 18) covers an isolated group of mountains about 16 miles northeast of Currie, a town on the Nevada Northern Railway. The northern end of the district was named Mizpah in 1905 on the discovery of the veins about 6 miles east of Mizpah siding on the railroad. Most of the properties in the southern part of the district are on the eastern flank of the mountains, extending from the summit eastward beyond Spring Canyon. (See fig. 6.)

All the properties are easily reached by good roads from Currie, though better roads for heavy traffic from the mines on the east side of the mountains reach the railroad at Dolly Varden siding by way of Antelope Valley.

There are many good springs on the east side of the range, and the principal settlement is Moore's camp at Watson Spring. Dolly Varden Spring, formerly called Last Chance, or Cloverdale station, is at the break of the mountains and Antelope Valley. There are small springs on the west side of the mountains in the vicinity of Mizpah and on the Currie road.

Throughout this group of hills water has been reached in most of the mine workings at about 60 feet.

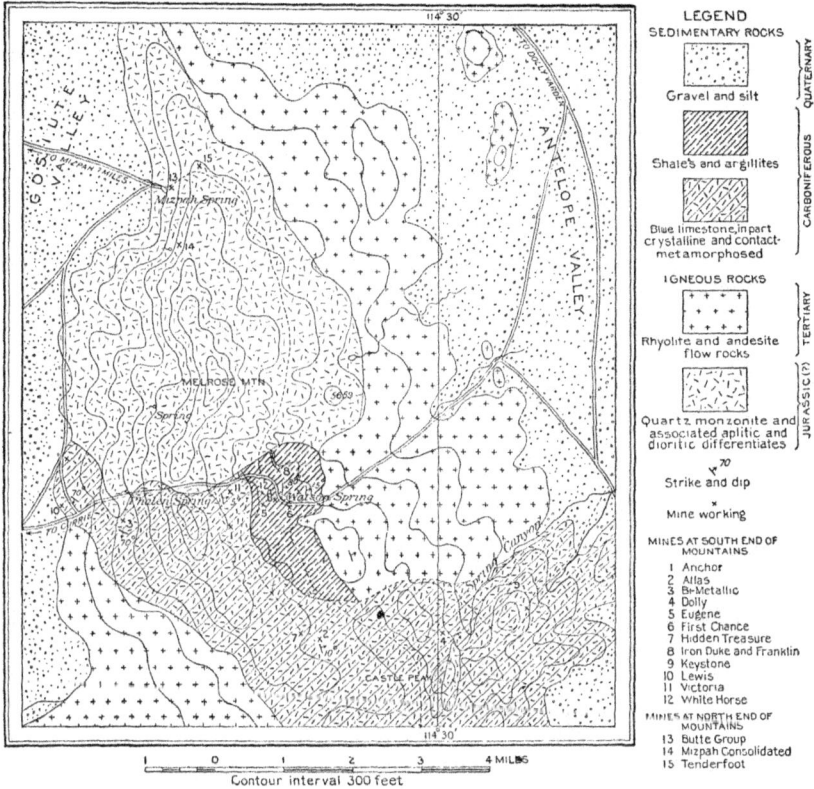

FIGURE 6.—Sketch map of the Dolly Varden (Mizpah) district, Elko County, Nev., showing reconnaissance geology, roads, springs, and principal properties.

TOPOGRAPHY.

The hills of the district, known to the geologists of the Fortieth Parallel Survey as the Wachoe Mountains,[1] rise abruptly 2,000 feet from the Gosiute Valley on the west. On the east they rise less steeply, sloping up at a moderate angle from Antelope Valley. The mountain mass is 8 to 9 miles long north and south and is 4 miles wide at its widest portion, near the south end. At the north end a long spur separates the Gosiute and Antelope valleys; and at the south a broad elevated table-land connects the Dolly Varden Mountains and the north end of the Schell Creek Range.

---

[1] Emmons, S. F., U. S. Geol. Expl. 40th Par. Rept., vol. 2, pp. 476–483, 1877.

GEOLOGY.

DOMINANT TYPES OF ROCK.

The largest part of this group of hills, and in fact all the higher summits north of Dolly Varden Pass, is composed of quartz monzonite. Most of the country rock south of the pass is sedimentary. Metamorphosed shales predominate in the vicinity of Watson Spring and limestones are exposed near the west base of the mountains along the road and in the vicinity of Castle Peak. (See fig. 6.) Later eruptive rocks form the lower hills on the east and south, overlying both the intrusive rock and the sediments.

SEDIMENTARY ROCKS.

The limestones that cover most of the southern portion of the district (see fig. 6) are medium thick bedded and are light blue where unmetamorphosed but in the vicinity of the ore deposits are white and crystalline and contain more or less lime silicate produced by the intrusion of the quartz monzonites. On the ridge west and south of the Hidden Treasure property they are probably less metamorphosed than elsewhere in the district. No fossils were found by the writer, but the geologists of the Fortieth Parallel Survey report finding *Rroductus subhorridus* and *Athyris roissyi*, both "Coal Measures" forms, in the limestones of Castle Peak.[1]

Lime shales, argillites, and' cherty shales, all more or less contact metamorphosed, overlie the limestones and underlie a small area in the vicinity of Watson Spring. (See fig. 6.)

INTRUSIVE IGNEOUS ROCKS.

Dolly Varden (Melrose) Mountain, is composed of a coarse-grained, somewhat porphyritic quartz monzonite. On weathered surfaces it is a deep brownish red but shows lighter-colored facies in many places. It is cut by a series of north-south joints which dips 80° W. and by an east-west series which stands vertical. A rather prominent sheeting dips 10°–15° E. This rock weathers in rounded forms but is very resistant.

The quartz monzonite is composed of oligoclase-andesine, microperthite, ferromagnesian minerals, and quartz, named in the order of their abundance. Green hornblende and brown biotite are everywhere present but in varying proportions. Apatite and zircon are abundant, the former often in large crystals. The rock is inequigranular, and, aside from the accessory minerals, few of its constituents show crystal faces. The quartz occurs as small interstitial

---

[1] Emmons, S. F., op. cit., p. 478.

fillings between crystals of the other constituents and is not conspicuous even in thin sections.

The geologists of the Fortieth Parallel Survey remarked the peculiarities of this "granite" and assigned it to the Jurassic,[1] because of its resemblance to other bodies in western Nevada. The following analysis is taken from their report:

Analysis of quartz monzonite from the Dolly Varden (Mizpah) district.

[T. M. Drown, Lafayette College, Pa., analyst.]

Silica	55.53
Alumina	18.65
Ferrous iron	6.14
Manganous oxide	.17
Lime	5.62
Magnesia	3.37
Soda	4.84
Potassa	5.20
Ignition	.65
	100.17

The edge facies of the stock are of finer grain and generally carry a larger proportion of dark minerals than the main mass of intrusive rock. In not a few places bodies of the quartz monzonite consist of minerals that are of pegmatitic size.

The quartz monzonite is clearly intrusive into limestones of Carboniferous age but has caused relatively little contact metamorphism in proportion to its size. Narrow zones and lenses of lime silicate rock appear at many places along the southern contact, particularly in the area underlain by limestone. Green garnet, biotite, epidote, and tremolite are the chief contact-metamorphic minerals and are in some places associated with limonitic oxidized copper ores that are evidently derived from chalcopyrite and pyrite originally deposited with the lime silicate minerals.

In the stock there are two types of dike rocks, both of which are probably derivatives of the quartz monzonite magma, though carrying the constituent minerals in differing proportions. One set of aplitic dikes in the Mizpah district is composed almost exclusively of oligoclase, microperthite, orthoclase, and quartz. Another series of dikes in the same vicinity are fine-grained dark-gray rocks consisting of orthoclase, plagioclase, biotite, hornblende, and quartz. In both of these types apatite and zircon are abundant accessory minerals, as they are in the quartz monzonite. The basic type of dike was noted by the geologists of the Fortieth Parallel Survey,[2] who mapped an area of diorite north of Melrose Peak.

---

[1] Emmons, S. F., op. cit., pp. 477–478.   [2] Idem, p. 478.

The low hills on the eastern side of the Melrose (Dolly Varden) Mountain are carved from flow rocks of two types. Rhyolites, glassy to almost wholly crystalline, in a multitude of brilliant colors cover by far the larger part of the area shown on figure 6. They are apparently younger than the dark andesites that outcrop in a few places where the rhyolites have been eroded. As there are, so far as could be learned, no ore deposits in the belt of extrusive rocks, little time was spent in examining the country they cover. Where seen the flows all have a decided though low dip to the east. In the vicinity of Watson Springs the rhyolite is a rather dark brown porphyritic rock, with phenocrysts of plagioclase feldspar set in a partly glassy matrix in which orthoclase, quartz, and biotite can be determined.

South of the mountains the elevated table-land is formed of nearly horizontal flows of Tertiary lavas, which at a distance are dull brownish in color. This region was not crossed by the writer, except in the vicinity of Currie, where the rocks are glassy rhyolites.

MINERAL DEPOSITS.

HISTORY AND PRODUCTION.

So far as could be learned, the first discoveries of mineral in the Melrose (Dolly Varden) Mountain were made in 1869, at the silver-lead mines in the southeastern hills, near what was called Hicks Spring, later Last Chance Spring, and now locally Dolly Varden Spring. In 1872 the copper ore at the Victoria was opened and for about two years was actively worked, the ores being smelted in a Mexican furnace at Dolly Varden Spring. Since the closing of the Victoria mine development in the district has been slight. Some excitement was caused by the discovery of the gold-bearing veins in the Mizpah section in 1905, but it was apparently short lived. Small shipments of ore have been made from time to time from the Victoria dump, and a little lead-silver ore has been extracted from the mines east of Castle Peak. In September, 1913, there were about ten prospectors in the mountains. One company had been operating a churn drill for a couple of years, but at the time of visit it was understood that the work was to be discontinued.

The production from the Victoria, Keystone, Eugene, and Hidden Treasure mines in the early days can not be learned, but it appears not to have been large.

Since 1908 the production, as shown in the following table, taken from the Mineral Resources reports published annually by the United States Geological Survey, has not been large. Probably most of the ore was shipped in small lots from various properties.

Production of the Dolly Varden district, 1908–1912.

Year.	Gold.	Silver.	Copper.	Lead.	Total value.
	Fine ounces.	*Fine ounces.*	*Pounds.*	*Pounds.*	
1908		545	35,272	9,071	$5,326
1909	9.92	3,750			3,031
1910	5.85	3,365	776	53,630	4,396
1911	7.84	1,768	8,171	32,369	3,577
1912	.92	540	24,667		4,421
	24.53	9,968	68,886	95,070	20,751

CHARACTER OF DEPOSITS.

The deposits of the Dolly Varden district are clearly related to the quartz monzonite intrusive. They occur as contact-metamorphic and replacement deposits in the area of Carboniferous limestones and shales at the south end of the mountains and as veins in the intrusive rock in the vicinity of Mizpah.

Deposits in the sediments.—The deposits in the sedimentary area are of two well-defined types. Oxidized copper ores, presumably derived from original deposits of pyrite and chalcopyrite, are usually associated with the lime silicate minerals and are nearer the intrusive contact than the lead deposits. The latter, found only in the extreme southeast part of the district, are typical replacements, both as fissure and bedded, occurring near fissures in limestone.

Copper deposits.—The ore minerals of the copper deposits so far developed are all of the oxidized variety, though kernels of the original chalcopyrite and pyrite remain at many places, and at not a few localities, where development has penetrated below water level, slightly cupriferous pyrite is seen to be the chief original sulphide. Chrysocolla, copper pitch ore, and malachite are most abundant, though chalcocite can usually be seen in the richer specimens of ore and bornite is exceptionally present. Limonite, as is to be expected, is present in abundance in all the copper deposits. The gangue minerals are quartz, green garnet, light-green biotite, and tremolite (with calcite and actinolite at some localities). The metamorphism due to the intrusion of the quartz monzonite magma has not been as intense as in some districts, and the croppings of the lime silicate zones and ore bodies are few and small.

It seems rather doubtful if any large bodies of enriched chalcocite ore will be found in this area, as the water table is near the surface— 90 feet at Moore's shaft and 60 feet at the Anchor mine. At Moore's shaft the zone of chalcocite ore has apparently been traversed and is reported to be very narrow. At the Anchor mine the shaft is now at the water level, but the chalcocite zone has not yet been reached.

42680°—Bull. 648—16——6

The original ore apparently contains only a small proportion of copper-bearing minerals. The same remarks are pertinent to the deposits of the Iron Duke and Franklin group and to those of prospects a short distance west of Watson Spring.

The argentiferous lead-bearing replacement deposits in the vicinity of Castle Peak are closely associated with north-south fractures cutting the limestones, along which the ore-bearing solutions have evidently moved. The principal minerals of these deposits, all of which are silver bearing, are cerusite, anglesite, and residual kernels of galena. Here and there a little copper carbonate stain is seen.

Deposits in quartz monzonite.—Small quartz veins characterized by chalcopyrite and minor pyrite and bismuthinite are typical of the veins in the intrusive rock. They are gold bearing, and some free gold is said to occur in the veins near Mizpah Spring. The solutions which deposited the veins have also altered the adjacent quartz monzonite, though usually in very narrow zones. Few such bands of sericitized and calcitized rock are over a foot in width, and most of them measure only a few inches.

At one place one-half mile east of Mizpah Spring an irregularly shaped body of thoroughly altered quartz monzonite about 300 by 200 feet in maximum dimensions somewhat resembles the croppings of the upper leached portion of the "porphyry copper" deposits.

THE PROPERTIES.

SOUTH END OF THE MOUNTAINS.

Anchor mine (No. 1, fig. 6).—The Anchor property, about $1\frac{1}{2}$ miles southwest of Watson Spring, belongs to W. G. White and is developed by a 60-foot incline equipped with a California whim. The incline dips 60° SE. along a zone of crushed mineralized limestone which strikes N. 55° E. This zone is 10 feet wide and carries abundant disseminated pyrite throughout that width. A little copper stain is seen through the mass. Water has been found at a depth of 60 feet, and it seems possible that below this level there may be a narrow zone of chalcocite ore. On the surface a series of low cliffs and saddles on a north-south strike seems to indicate a fracture that cuts across the country just west of the shaft.

Atlas claims (No. 2, fig. 6).—The Atlas claims, belonging to Thomas Boofer, lie east of the Hidden Treasure, in comparatively level ground, underlain by yellowish lime shales and limestones which dip 10° to 15° E. Parallel to the bedding there are small replacement lenses and veinlike deposits of chalcopyrite and pyrite which are partly altered to copper pitch ore and malachite.

Bimetallic group (No. 3, fig. 6).—The Bimetallic claims, belonging to George Gordon, are located along the contact of limestone and quartz monzonite on the west side of the mountains south of Dolly Varden Canyon and Phalen Spring. The quartz monzonite near the contact is of finer grain than in the main body of the stock; is light gray; and, as seen in thin sections under the microscope, consists of andesine, a microperthitic intergrowth of plagioclase and orthoclase, green hornblende, quartz, and brown biotite. Zircon and apatite are abundant accessory minerals, and some magnetite is present. At some places biotite is more abundant than hornblende; at others the reverse is the case. Narrow zones of lime silicate minerals are present at many places near the contact. Green garnet, epidote, and plates of light-green mica 1 inch in maximum diameter are the most abundant contact minerals, though some tremolite and iron oxide are commonly present. At numerous places on these claims limonitic copper carbonate ores are associated with lime silicate zones. Residual pyrite and chalcopyrite are occasionally seen, and some chalcocite is present, enveloping the original sulphides.

The principal development work is a 265-foot tunnel, which runs north along the contact on the east side of a tongue of quartz monzonite, exposing many small deposits of copper carbonates. On another claim a 65-foot inclined shaft is sunk in coarsely crystalline white limestone, the bedding of which strikes N. 70° W. and dips 65° S. Tremolite was noted with the copper carbonate ore on the dump.

Dolly mine (No. 4, fig. 6).—The Dolly mine, southeast of Spring Canyon, is one of the older lead-silver mines of the district. Most of the work at the property is said to have been done in the early seventies. The deeper workings are inaccessible at present, but in the exposures in the surface cuts the oxidized lead ores are seen to occur as irregular replacements parallel to the bedding of white limestone as well as along a north-south fissure which dips 75° W. The limestones strike east-northeast and dip south at low angles.

The ores consist of cerusite and anglesite, in places associated with a little copper carbonate and everywhere with considerable limonite. No galena was noted in the small amount of ore remaining about the old workings.

On the Dolly group, 300 feet north of the Murphy shaft, a drill hole was sunk to a reported depth of 240 feet. It starts in limestone, and, to judge from the sludge piles, continues in that formation throughout. Some pyritic sludge is said to have come from a depth of 85 feet down to the bottom of the hole. Some black shaly material, said to have come from a depth of 80 feet, is reported to carry silver. However, qualitative chemical tests made in the Survey laboratory failed to show the presence of that metal.

Eugene patent (No. 5, fig. 6).—The Eugene patent, one-half mile west of Watson Spring, is the property of the Victoria United Copper Mining Co. The abandoned workings on the ridge south of Watson Canyon are on a large cropping of brown lime silicate rock, mostly garnet, quartz, and limonite, in which there are some small bodies of high-grade copper carbonate ore. Chrysocolla and malachite are the principal ore minerals, to judge from the ore on the dump, though some chalcocite and a little chalcopyrite were noted.

First Chance group (No. 6, fig. 6).—The First Chance group of 18 claims, covering both sides of Watson Canyon east of Watson Spring, is the property of J. L. Moore, whose camp is at Watson Spring. These claims are largely underlain by argillites and lime shales, though in the eastern part of the group the rhyolite flow rocks are found on the ridges. The shales strike a few degrees west of north and dip eastward at low angles except where locally disturbed—as at Watson Spring, where east of what appears to be a minor fault they dip 80° E. Locally there are small areas of lime silicate minerals, though metamorphism is not so strong on these claims as on those nearer the contact. At a number of places copper carbonates are seen associated with iron oxides and the contact-metamorphic minerals.

The principal development work, a vertical shaft said to be 260 feet deep, in which water stands 90 feet below the collar, is about one-fourth mile east of Watson Spring. The dump shows much coarsely crystallized pyrite, said to have come from a depth of 70 feet, which is tarnished with the peacock colors of copper. Above 70 feet the shaft is all in oxidized limonitic material. Some specimens of chalcocite, occurring as small veinlets in the pyritic material, are said to have come from a narrow belt at 150 feet. Below this chalcocite zone the shaft is said to be again in argillites and limestones, in which there are bodies of somewhat tarnished pyrite. It is proposed to continue this shaft either by sinking or drilling. It seems fairly certain that the zone of enrichment has been pierced at 150 feet, where the chalcocite was found, particularly as permanent water now stands above this level.

Hidden Treasure claim (No. 7, fig. 6).—The Hidden Treasure claim of the Victoria United Copper Co. is 3 miles south of Watson Spring. The deepest work is a 30-foot shaft. Ore is said to have been taken from the surface cuts on the ledge. The croppings, deeply stained with iron and very siliceous, stand well above the surrounding limestone. Copper carbonates are rather abundantly distributed through this mass, which can be traced for over a claim's length north and south along the strike.

Iron Duke and Franklin group (No. 8, fig. 6).—The Iron Duke and Franklin group of eight claims, the property of George Gordon, is about half a mile north-northwest of Watson Spring, and a quarter of a mile southeast of the sedimentary quartz monzonite contact. In this vicinity thin-bedded shaly limestones and cherty lime shales, the prevailing rocks, dip to the east at low angles and apparently overlie the limestones exposed farther west near the summit of the range. These beds have been intruded by a series of northeast-trending dikes of pinkish-gray feldspathic porphyry. Thin sections of this type of rock, examined microscopically, show andesine-labradorite and orthoclase phenocrysts in a microgranular groundmass composed chiefly of sericitized feldspar with minor amounts of quartz filling interstices. What appears to have been hornblende is altered to chlorite, iron oxide, and carbonate. Apatite is very abundant. The rock closely resembles some facies of the rock of the main stock, and there is little question that these dikes are off-shoots of the larger mass.

Along and near the dikes are narrow zones of epidotized pyrite-bearing rock, which are more abundant in the cherty lime shales than in the limestones. Besides epidote and pyrite, the dikes contain some calcite, quartz, and feldspar, as shown by thin sections. Bodies of these minerals have been developed at a number of places in the group. The pyrite is slightly cupriferous, but so far as noted only very small amounts of copper minerals are present in any deposits as yet opened. The main tunnel runs east 200 feet through altered sediments and crushed and leached quartz monzonite porphyry. A little pyrite and chalcopyrite ʹ disseminated in the latter rock, which is somewhat iron stained.

Keystone mine (No. 9, fig. 6).—The old Keystone mine, from which it is said that some lead-silver ore was taken in the early days, is on the limestone ridge southeast of Spring Canyon. The old, inaccessible workings are said to be rather short. Reports indicate that the argentiferous anglesite-cerusite ores occurred as replacement bodies parallel to the bedding near north-south fissures which dip 45° to 50° E.

Lewis group (No. 10, fig. 6).—The Lewis group of six claims is on the flat northwest of the mouth of Dolly Varden Canyon. The principal development work, an inaccessible inclined shaft, is a few rods north of the main road in the canyon. This shaft, said to be 250 feet deep, follows the bedding of white, coarsely crystalline limestone which strikes N. 70° W. and dips 40° S. It is about a quarter of a mile west of the contact of limestone and quartz monzonite and some contact-metamorphic minerals, chiefly tremolite and actinolite, occurring in small radial masses, were found in the ore on the dump. This ore is largely oxidized, carrying limonite, chryso-colla, and minor malachite. Specks of chalcopyrite, surrounded by a dark sooty mineral, probably chalcocite, are seen in the better grade of material. At several cuts and pits on other claims oxidized copper ores are associated with the contact minerals.

Victoria claim (No. 11, fig. 6).—The patented Victoria claim is on the east side of the mountains about halfway between the summit and Watson Spring in the south fork of Watson Canyon. The croppings, iron-stained cellular quartz and jasper, with some copper carbonates, stand about 10 feet above the surrounding country and cover an area about 100 feet long north and south by 75 feet wide. The country rock in this vicinity is limestone, and a rather indistinct bedding appears to dip 5° E. The shaft, equipped with a small steam hoist, is under water below a depth of 80 feet, and could not be entered. The limonitic ore on the dump is cellular and very siliceous, consisting of brecciated contact-metamorphosed limestone and quartz, with chrysocolla, malachite, and a small amount of chalcocite. Tremolite, the principal contact mineral of the deposit, occurs as veinlets of fibrous asbestos-like material and in fibers throughout yellow siliceous altered limestone. It is said that very little high-grade ore remains on the dump, from which several shipments of sorted ore have been made.

White Horse group (No. 12, fig. 6).—The five claims of the White Horse group, belonging to G. H. Gordon and F. A. Herreld, are on the ridge south of Watson Canyon about three-fourths of a mile west of the Watson Spring. Several croppings of quartz-garnet rocks carrying iron oxide and oxidized copper ores are exposed in the shallow workings.

Other prospects.—A few hundred feet west of Watson Spring there are some shallow workings on a body of iron-stained pyrite-bearing lime silicate rock near a dike of leached, somewhat pyritized quartz monzonite porphyry. Small areas stained with copper carbonate indicate the presence of a little copper in the original ore which, however, appears to consist entirely of pyrite.

Some lead-silver ores occur near the Keystone Mine on two groups of claims controlled by G. H. Gordon and Capt. Mahan. A little development work has been done on both groups of claims and a small quantity of argentiferous lead carbonate ore has been shipped.

A few prospects in the limestone area north of the Lewis ground, on claims at the west base of the mountains, belong to Chris Krez. The deposits are similar to those on the Lewis ground.

NORTH END OF MOUNTAINS.

Butte group (No. 13, fig. 6).—Mr. T. F. Banigan, of Currie, Nev., controls the Butte group of 12 claims in the hills immediately east of Mizpah Spring, where the gold discoveries were made in 1905. Several of these claims have shallow shafts and pits, the deepest of which is about 20 feet. With one exception the deposits are in veins ranging from a few inches to 2 feet in width and striking N. 16°–27° W. with moderately steep west dip. They cut a reddish-gray medium-grained quartz monzonite porphyry and are later than a series of quartz diorite and aplitic dikes. They consist of quartz, chalcopyrite, pyrite, and bismuthinite, and are said to carry free gold. Bornite, chalcocite, copper pitch ore, malachite, and azurite were seen as alteration products and, in the ores so far developed, are probably more abundant than the original sulphides. Some of these veins are frozen, but postmineral movement has produced a gouge along many of them. The quartz monzonite near the veins is bleached, and sericite is usually developed from the feldspars. Fragments of the altered rock are inclosed in the veins at a few places.

The Red Hill claim, named from its mass of iron-stained leached porphyry, is cut by numerous fractures, none of which appear to contain veins. Instead, the igneous rock is altered to a soft light-gray mass of quartz, calcite, and sericite, stained with iron and in some places showing specks of what appears to be chalcocite. It is said that low assays for copper have been obtained from samples taken at various places over an area 300 feet long by 200 feet wide. In a chemical test made on some of this material collected by the writer the presence of very small amounts of copper was shown. Most of this leached body does not appear particularly promising as an ore, though at a few places along the fracture zones there are fair indications of copper. It, however, may represent the leached croppings of a small "porphyry copper" deposit.

Mizpah Consolidated group (No. 14, fig. 6).—The Mizpah Consolidated Gold & Copper Mining Co., of Philadelphia, owns a group of 21 claims near the summit of the north spur of the range 1½ miles southeast of Mizpah Spring. Several small veins similar to that at Mizpah have been exposed on the surface by the location work on the different claims. The principal development work is a 1,300-foot crosscut tunnel, which runs N. 72° E. through the quartz monzonite. It is near the summit and can hardly attain a greater depth than 300 feet. In the length of the tunnel 10 small quartz stringers have been cut, a few of which are barren but most of which carry a little pyrite and chalcopyrite. Sericitization of the quartz monzonite has taken place along all these fissures. About 5 feet of soft pyrite-impregnated bleached quartz monzonite at the face of the tunnel seems to indicate that a vein or fissure is not far distant. This tunnel was driven to cut what is known as the Northern vein, said to be 18 inches to 2 feet wide and to contain some very good grade copper-gold ore at the surface. It dips 45° W. and from surveys would seem to be very near the vein.

Tenderfoot shaft (No. 15, fig. 6).—The Tenderfoot shaft is on the summit of the ridge at the north end of the Dolly Varden Mountains about a mile east of Mizpah Spring. It could not be entered, as water stands 100 feet below the collar in the 120-foot incline. It is sunk on a narrow vein, which strikes N. 40° W. and which dips 60° SW. near the surface and 45° lower down. The ore on the dump indicates that the vein is nowhere over 3 inches wide (average about 2 inches), and is frozen to the sericitized pyrite-impregnated quartz monzonite walls. The alteration of the wall rocks is apparently limited to a few inches on either side of the vein. Chalcopyrite and minor amounts of pyrite with both white and smoky quartz constitute the ore. In most of the ore the sulphides occupy the medial portion of the vein with quartz next the walls, but this is not everywhere the case. All of the ore is partly oxidized, copper pitch ore being the usual product, with sometimes a little malachite. It is said that the ore carried some bismuth, but the blowpipe failed to show it in any of the writer's tests on material collected by him.

Other prospects.—Half a mile south of Mizpah Spring there are a few prospect pits on a vein which strikes N. 20° W. and dips 85° W. This vein is 2 to 4 inches wide and is frozen to the quartz monzonite walls. It carries quartz and remnants of chalcopyrite, chrysocolla, malachite, and a little cuprite.

KINSLEY DISTRICT.

LOCATION AND ACCESSIBILITY.

The Kinsley district (No. 6, fig. 1, p. 18) is at the south end of a peculiar long, narrow ridge which lies southeast of the Dolly Varden Mountains on the west side of Antelope Valley. (See Pl. I.) Kinsley Spring, on the southwest side of this ridge (see fig. 7), is 38 miles southwest of Currie, the shipping point. The mining properties are all grouped at the south end of the mountains about a stock of quartz monzonite porphyry. So far as could be learned they are all in Elko County, though very near the county line, which is probably not so far from the mouth of Kinsley Canyon, on the road to Tippet. The mines are in secs. 13 and 14, T. 26 N., R. 67 E., and secs. 18 and 19, T. 26 N., R. 68 E.

ECONOMIC CONDITIONS.

Water is scarce in the hills, there being no springs except at Kinsley and on the road to Currie, about 8 miles northwest of the camp. Kinsley Springs flows about 8 barrels of highly mineralized water in 24 hours. It issues from red andesite near the contact with the

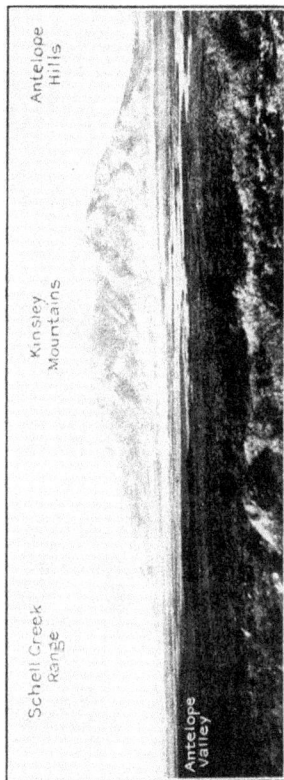

A. KINSLEY MOUNTAIN.

Looking S. 30° W. across Antelope Valley from White Horse.

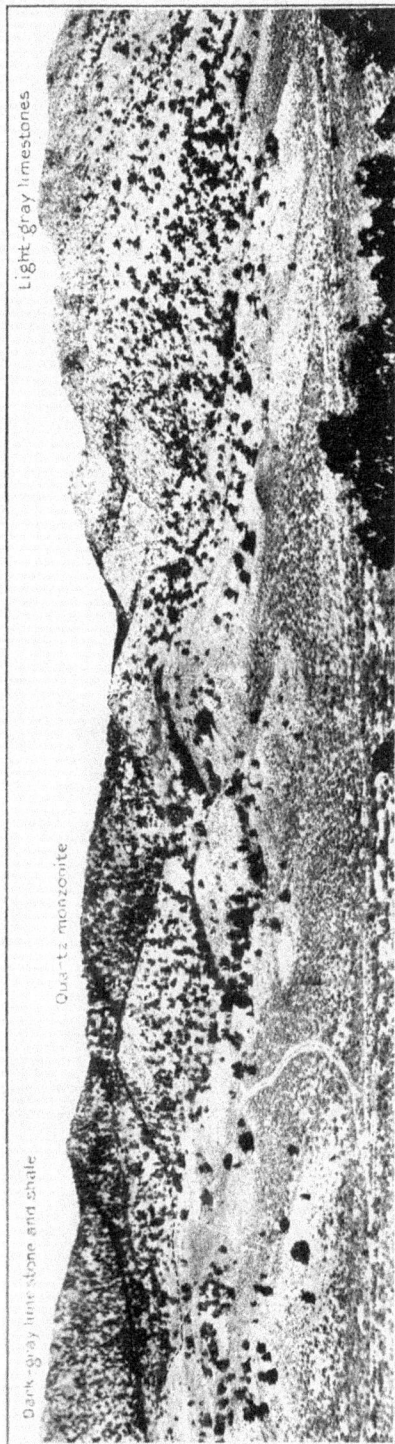

B. WEST-SOUTHWEST PART OF KINSLEY DISTRICT.

Showing contact of the quartz monzonite intrusive mass and the older sedimentary formations, looking east-southeast from a point about half a mile south of Kinsley Spring.

underlying limestone. It is said that Boone Spring flows a somewhat larger amount of water, but that it is more heavily mineralized than Kinsley.

There is a fair growth of piñon and juniper on Kinsley Mountain, and in the granitic area some fair-sized trees are found. Desert grasses and forage shrubs are abundant in the valleys and in Antelope Valley are said to furnish a fine winter range for sheep and cattle.

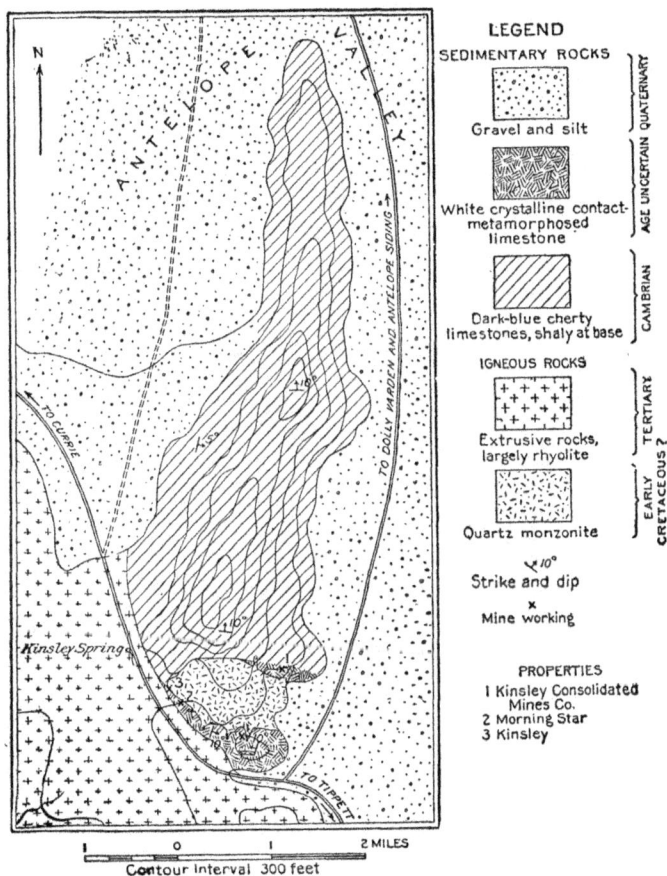

FIGURE 7.—Sketch map of the Kinsley district, Elko County, Nev., showing reconnaissance geology and the positions of the principal prospects.

TOPOGRAPHY.

Kinsley Mountain (fig. 7) is a long narrow tongue or island which rises nearly 2,000 feet above the level flat of Antelope Valley. (See Pl. IV, A.) To the south and west low volcanic hills stretch to the Antelope and Schell Creek ranges, several miles distant.

As seen from the northwest, this narrow spur looks like a rugged wedge dividing Antelope Valley. The slopes are steep, and near the center and top of the mountains there are many shear escarpments formed by what appear to be massive beds of limestone. In the vicinity of the mines the topographic features are in general low and rounded, contrasting sharply with the summit of the mountains. (See Pl. IV, *B*.) In detail, however, these apparently smooth hills are rough and are formed by numerous layers of limestone. The intrusive rock weathers in rounded forms and has been eroded more easily and rapidly than the surrounding sediments.

GEOLOGY.

SEDIMENTARY ROCKS.

All of the northern part of Kinsley Mountain is composed of dark-gray sedimentary rocks, which from a distance appear to dip about 10° N., modified along the west side of the ridge by an eastward dip. Near the center of the mountains a break seems to cut the ridge. (See Pl. IV, *A*.) The ridge was not studied in detail, but prospectors report that it contains no intrusive rocks except the stock near the mines.

The geologists of the Fortieth Parallel Survey report[1] that a rather complex series of alternating igneous and sedimentary rocks, which they considered Archean, occur on the side of the mountains, but as they may have visited only the southern part of the group, it seems rather doubtful if they interpreted the conditions correctly, especially as the quartz monzonite is clearly intrusive into the sediments.

Among a few fossils collected by the writer from thin-bedded shales and shaly limestones just north of the contact at the summit of the ridge, Edwin Kirk, of the United States Geological Survey, found some remnants of "indeterminable trilobite fragments," which he considered to indicate a Cambrian age. Overlying the shaly beds there are about 300 feet of nearly black, dense limestones in beds 3 to 5 feet thick, which become lighter colored and more massive at the top and finally grade into a light bluish gray.

In the vicinity of the quartz monzonite stock the limestones are crystalline and light colored, doubtless owing to alteration at the time of the intrusion. These rocks are shown by a separate symbol on figure 7, for their age is uncertain and they may be younger than the limestones of the north ridge, though it is possible that they are of the same age as the upper part of the series represented there.

Antelope Valley is filled with a great accumulation of gravels and sands and contains several playa basins. (See Pl. IV, *A*.) The alluvial cones about Kinsley Mountain are small but are fairly well developed.

---

[1] Emmons, S. F., U. S. Geol. Expl. 40th Par. Rept., vol. 2, pp. 483–485, 1877.

IGNEOUS ROCKS.

Extrusive rocks.—South and west of Kinsley Mountain a large area is underlain by flow rocks ranging from gray to brick red in color and from porphyritic to glassy in texture. These rocks were called rhyolites by the geologists of the Fortieth Parallel Survey.[1]

In the vicinity of Kinsley Canyon, however, andesitic flows prevail. Near the mouth of the canyon a gray porphyritic andesite contains phenocrysts of labradorite and numberless minute laths of plagioclase in a partially glassy groundmass. Magnetite is abundant in the groundmass, and what appears to have been hornblende is now altered to a reddish iron oxide. Near Kinsley Spring the brownish-black red-weathering andesites are more glassy. A few phenocrysts of andesine and augite can be recognized in thin sections, and spots of iron oxide in the form of hornblende seem to indicate the former presence of that mineral. A few plagioclase microlites are irregularly distributed through the brownish glass which forms the groundmass.

Intrusive rocks.—The intrusive stock at the south end of Kinsley Mountain is a quartz monzonite of rather coarse grain and in places of somewhat porphyritic texture. It is light gray when fresh, but weathers to tawny and brown. The constituent minerals, named in the order of their abundance, are orthoclase, oligoclase-andesine, quartz, brownish-green biotite, and green hornblende, all of megascopic size. Apatite, magnetite, and titanite are fairly abundant accessories. There is little difference in either texture or composition between the edge facies and the center of the stock, but the offshoot dikes are as a rule rather free from the dark minerals. Aplitic and basic differentiates are not abundant, though both types of later dikes were seen at a few places near the center of the stock. The basic dikes are usually porphyritic and of fairly coarse grain. The phenocrystic, zonally built plagioclase feldspars are set in an inequigranular groundmass of hornblende, biotite, quartz, orthoclase, and plagioclase. The aplitic dikes are fine-grained light tawny rocks composed essentially of quartz and the two varieties of feldspars.

ALTERATION.

The dolomitic limestones on all sides of this stock have suffered some contact metamorphism. Few of the belts of lime silicate rock are over 10 feet wide and most of them are 2 to 4 feet in maximum width. They follow the bedding planes rather than the actual contact but are rarely far removed from the intrusive rock. Tremolite and wollastonite are the commonly developed contact minerals,

[1] Emmons, S. F., op. cit., p. 485.

though in places a light-green mica and rarely a gray garnet are present. Copper minerals and a little magnetite are associated with many of the lime silicate zones.

ORE DEPOSITS.

HISTORY AND DEVELOPMENT.

According to the State mineralogist of Nevada,[1] the copper deposits of the Kinsley district were discovered in December, 1862, by Felix O'Neil, who was driven out of the country by the Mormons in 1863. The veins were rediscovered in 1865 by George Kinsley, a soldier, for whom the district was named. Little work was done till 1867, when a Mexican furnace was built at the springs. This did not prove a success, and by 1872 the camp was largely abandoned. Comparatively little work has been done in the district. The most extensive development is on the Dunyon claims on the northeast contact, though the shaft at the Morning Star is said to have been over 200 feet deep, and the dump at the old Kinsley property indicates considerable excavation.

It was not found possible to obtain even estimates of the early production from this camp, and the United States Geological Survey has not published separate figures of its production in recent years. It is known that in 1913 at least 20 tons of ore, carrying 28.70 ounces silver per ton, 8.6 per cent lead, and 3.42 per cent copper were shipped from the dump of the Morning Star, and it is probable that several carloads of similar grade were shipped by the lessees operating the property during 1913.

CHARACTER OF DEPOSITS.

The developed ore deposits of the Kinsley district are all of contact-metamorphic origin, being clearly connected with the intrusive quartz monzonite and usually associated with tremolite and green biotite. They have as a rule a more or less veinlike form, for most of them lie along fault fissures which strike nearly north and south, though the Morning Star deposit strikes nearly east and west. Besides the contact-metamorphic minerals already mentioned, chrysocolla, malachite, azurite, and limonite are the principal metallic minerals, though remnants of the original pyrite and chalcopyrite are to be found in most of the workings. Copper pitch ore also occurs. A little chalcocite and more rarely bornite is present in some places.

---

[1] White, A. F., Nevada State Mineralogist Third Biennial Rept., for 1869–70, p. 63 [1871].

A few small quartz veinlets in the intrusive stock carry pyrite, chalcopyrite, and a small amount of galena and the alteration products usually derived from those minerals. These veins are frozen to the walls, which show very little alteration.

Kinsley Consolidated Mines Co. claims (No. 1, fig. 7).—The Kinsley Consolidated Mines Co. owns numerous claims on the northeast contact of the intrusive stock about 2½ miles east of Kinsley Spring. The largest workings are about one-eighth mile east of the main contact but near a dike of quartz monzonite which runs northeast from the stock. In an eastward branch of the lower tunnel the quartz monzonite dike is cut 60 feet from the mouth of the tunnel. A whim shaft, said to be 100 feet deep, and four tunnels constitute the work. The lower tunnel, whose mouth is approximately level with the collar of the shaft, is driven northwestward for 100 feet to intersect the vein, which is then followed N. 21° E. for 160 feet. The vein cuts white crystalline dolomite beds, which are essentially horizontal. It dips 55°–60° E. and varies from a tight barren fracture to a width of 2 feet, being more or less open throughout. In the wider portions of the vein at this level there is some cupriferous oxidized iron ore, both limonite and hematite, which is clearly an alteration product of cupriferous pyrite.

A 65-foot raise connects this tunnel with one above, whose mouth is 100 feet north of the shaft. In this tunnel the vein strikes N. 25° E. and stands nearly vertical above the tunnel but dips east below it. At its level the vein was apparently narrow and consisted largely of limonitic material with some lenslike bodies of good-grade oxidized copper ore, essentially chrysocolla, malachite, and chalcocite. Some replacement bodies parallel to the nearly horizontally bedded limestones occur along the fissure and appear to have contained copper ores associated with tremolite and light greenish-gray garnet.

Two tunnels lie between those described, the lower running N. 10° E. for 100 feet and exposing some small bedded deposits of copper carbonates and the higher running N. 20° E. for 290 feet. This last tunnel lies east of the fissure along which the ore solutions seem to have penetrated.

On the hills south of the stock near the mouth of Kinsley Canyon some prospects, presumably part of this company's holdings, have been opened on a N. 16° W. vertical vein in white crystalline dolomitic limestone, which strikes north and dips 10° to 15° W. The vein, which is 1 to 2 feet in width, consists of quartz, part of which is chalcedonic, calcite, limonite, and copper carbonates. Residual kernels of chalcopyrite are altered in part to bornite and chalcocite but largely to copper pitch ore, malachite, and chrysocolla. Limonite is abundant. This vein has been opened in numerous places along the strike for about 400 feet. A crosscut tunnel 75 feet below the croppings has been run 50 feet N. 78° W. without cutting the vein but has cut two small fissures which strike N. 5° W. and dip 85° E. and along which there are indications of mineralization. The tunnel and the main shaft on the vein are equipped with a crude tram, which lowers the ore nearly to the level of the road in Kinsley Canyon.

Morning Star shaft (No. 2, fig. 7).—The old Morning Star shaft is one-half mile east-southeast of Kinsley Spring. The claim belongs to the G. A. Lowe estate, but in September, 1913, was under lease to E. C. Rowland and John Fasano, who were engaged in cleaning out the shaft at the time of visit. They were also shipping the old dump, which is said to carry about $25 a ton in silver, lead, and copper. By September 30 they had opened the shaft, said to be 275 feet deep, to a depth of 30 feet, and had exposed the vein 25 feet east of the shaft on that level. The vein strikes N. 86° E. and dips 50° N. Next the dolomite hanging wall there is a foot of iron-stained gouge, below which there are 2 inches of limonitic copper carbonate ore, said to carry gold and silver. A second narrow streak of ore lies on the footwall separated from the hanging-wall streak by 2 feet of limestone. This ore seems to be a replacement along a faulted fissure, particularly in the brecciated material along the walls. Many fragments of unaltered dolomite are present in the ore, which consists of quartz, calcite, malachite, azurite, and chryso-colla. Some silver chloride was seen in the ore and no doubt is present with the blue and green copper stains that the prospectors speak of as silver chloride or bromide.

Kinsley mine (No. 3, fig. 7).—The old Kinsley surface workings, about one-eighth mile north of the Morning Star shaft, show some bodies of copper carbonate ores associated with contact-metamorphic minerals. The shaft is not accessible but is shown by the dump to be deep and to have apparently failed to intersect any ore bodies in the dolomite penetrated.

Other prospects.—At numerous places in the crystalline dolomite hills south of the stock there are shallow openings on small bodies of copper carbonate ores, which occur along nearly north-south fissures in the vicinity of quartz monzonite dikes. Tremolite and, more rarely, greenish mica are developed at a few of the contacts of these dikes, though usually the crystalline dolomite has not been further altered by metamorphism.

At one prospect about one-fourth mile southeast of the Morning Star a small amount of lead carbonate and some copper carbonates were seen along a north-south vertical fissure replacement deposit.

TOANO RANGE.

LOCATION.

The Toano Range extends southward from Toano Pass, east of Cobre (see Pl. I), for 65 miles, to the Elko County line. These mountains, called the Gosiute Range by the geologists of the Fortieth Parallel Survey,[1] trend very nearly true north and south, and stand as a narrow wall between the Great Salt Lake desert on the east and the Basin Mountain country of Nevada on the west.

TOPOGRAPHY.

The range is in few places more than 6 miles wide. It rises from the Salt Lake desert in a long gentle slope, broken near the summit by rough, irregular topography sculptured from the massive limestones and quartzites which form the main mountains. The western side forms a rough and precipitous wall, notched by deep, steep-sided canyons, along the eastern edge of Antelope and Gosiute valleys.

Silver Zone Pass, formerly called Middle Pass, about 16 miles south of Toano Pass, used by the Western Pacific Railway, is occupied by a body of what is probably intrusive granitic rock. Don Don Pass is a relatively low break through the range near its southern end, east of the Dolly Varden Mountains.

GEOLOGY.

Sedimentary rocks.—The range is composed for the most part of quartzites and limestones, the limestones prevailing. According to the geologists of the Fortieth Parallel Survey, the limestones reach their greatest development at Lookout Mountain, the highest peak, near the center of the range, a few miles northwest of Ferguson Spring. (See Pl. I.) This series of sediments was all mapped by the early geologists as Carboniferous, though it seems probable that the formations may include rocks ranging from Cambrian to the Carboniferous. A few fossils collected by the writer in 1913 at Ferguson Spring have been referred to the Mississippian by G. II. Girty, who found *Girtyina ventricosa* and *Echinocrinus* sp. among them. These fossils were taken from the upper beds of the eastern limb of the anticlinal fold south of Lookout Peak, and the immense development of limestones involved in this anticline makes it prob-

[1] Hague, Arnold, U. S. Geol. Expl. 40th Par. Rept., vol. 2, pp. 502–505, 1877.

able that a large part of the lower beds are older than the Mississippian.

Igneous rocks.—At least two bodies of granitic rocks occur in the range. The larger body, at Middle Pass, is described by the geologists of the Fortieth Parallel Survey as a coarse-grained [1] light-gray intrusive granite but should possibly be correlated with the quartz monzonitic intrusives common in eastern Nevada. South of Mount Pisgah, about opposite the north end of the Kinsley Mountains (see Pl. I), a small stock of typical quartz monzonite centers about White Horse (Zilig) Spring. The rock is a rather coarse grained porphyritic quartz monzonite with large phenocrysts of pink orthoclase and a few of quartz and biotite set in an inequigranular groundmass of oligoclase-andesine, orthoclase, quartz, and biotite, named in the order of their abundance. Apatite and magnetite are rather abundant accessories, and in places a little titanite is present. The edge phase of this stock has a finer grain, is not porphyritic, and carries some hornblende as well as biotite.

In the vicinity of Ferguson Spring there is a small area of light-colored rhyolitic or latitic flows. The rocks are cryptocrystalline and are so much altered that the microscopic study of thin sections yields little knowledge that can not be gained from hand specimens. Plagioclase and biotite are the phenocrystic minerals set in a very fine felsitic groundmass that contains some free silica.

About 4 miles south of Don Don Pass on the east side of the Toano Range the low eastward slopes are covered in a few places by thin flows of light to dark-gray rocks similar to those near Ferguson Spring. At one place the base of a gray, somewhat glassy porphyry was composed of a nearly black glass with a few phenocrysts of hornblende and feldspar.

ORE DEPOSITS.

LURAY DISTRICT.

About $4\frac{1}{2}$ miles east of Luray siding, of the Southern Pacific, on the south side of Toano Pass (No. 7, fig. 1, p. 18), there are a few undeveloped prospects on the Silver Star group and the property of the Cobre Lead Co. The deposits, which were not visited, are said to be in light and dark limestones which are somewhat folded but which in general strike north and south and dip 30° W. The dark limestones are said to underlie the nearly white crystalline limestones in which most of the ore occurs. Some of the oxidized ore from the Silver Star group, seen at Cobre, consists of white crystalline limestone and white opaline quartz. Chrysocolla, mala-

---

[1] Hague, Arnold, op. cit., p. 503.

chite, and some azurite occur both surrounding and in veinlets through a nearly black resinous substance which contains copper, lead, and iron, combined as oxide, carbonate, and sulphide. It is probably one of the isomorphous gel ore series, called copper pitch ore.

FERGUSON SPRING DISTRICT.

Geology.—On the west side of the range at Don Don Pass the older limestones are dark blue and are seamed with white calcite veinlets. They are thin bedded, becoming thicker toward the top. Over these beds are massive light blue-gray limestones, in which some Mississippian fossils were found at Ferguson Spring. The structure is rather complex, but in general this part of the range seems to be an unsymmetrical anticline with steep-dipping beds on the west and lower dips on the east. At Don Don Pass the western limb has been eroded so that the beds all dip eastward, as if that part of the range were a monoclinal fault block. In the vicinity of Ferguson Spring (No. 5, fig. 1, p. 18) a transverse anticlinal fold with a N. 40° W. strike seems to extend across the range. Ferguson Spring rises from the western limb of this anticline, where it is cut by an east-west break. Near the spring the brownish-gray fossiliferous limestones are somewhat cherty. The chert is dark and occurs in large masses in a few beds but not throughout all the limestones.

Mineral deposits.—The deposits near Ferguson Spring are said to have been known in the late eighties, but the oldest location notice, which was found on the Red Boy or Danger Line claim, was dated January 1, 1910. The name Danger Line was repeated on another location notice, dated April 27, 1912, found in the same monument. The notices call the district "Allegheny."

The deepest work is a 50-foot crosscut tunnel which undercuts one of the larger iron-stained croppings. At the depth of the tunnel there seems to be very little mineralization, but it may be that some ore will be found farther on, as the fissure zone has not been reached by the tunnel. Other places in the neighborhood show shallow pits and open cuts.

The ore bodies occur on the southwest side of a low hogback which rises south and west of Ferguson Spring and in the limestones immediately to the north. All the ores so far developed are oxidized and consist for the most part of limonite and barite in irregular replacement deposits that parallel the bedding along east-west fractures. The principal break along which mineralization has occurred seems to be parallel to but south of that on which the spring rises. The largest body is about 300 feet long and 30 feet wide, the croppings standing well above the surface. Some areas of copper carbonates occur in the limonite, but they are very small and few. Similar deposits are said to occur in a low hogback 3 miles east of the spring, on the northern limb of the same anticline.

On the whole the mineralization of this district seems to be weak, and it is rather doubtful if any commercial deposits will be found in the immediate vicinity of Ferguson Spring.

WHITE HORSE DISTRICT.

In the quartz monzonite stock, which occupies an area about 2 miles in diameter, forming the peaks on the southwest flank of Mount Pisgah (see Pl. I), there are some small veins. The monzonite is cut by north-south joints which dip 70°–80° W., by west-southwest joints which stand vertical, and by a flat eastward-dipping sheeting. The veins, which are usually parallel to the north-south sheeting, are composed of quartz and, so far as developed, carry only oxidized ores, showing much limonite, with a minor amount of copper and lead carbonates. The walls are sericitized for short distances on both sides of the veins. At the dump of the deepest shaft, which reaches a depth of 50 feet, there is a little residual pyrite and chalcopyrite, surrounded by copper pitch ore, limonite, and copper carbonates. A 120-foot tunnel northeast of Zelig Spring exposes two small veins.

FERBER DISTRICT.

Location of mines.—The Ferber district is in some low foothills 3 miles east of the main Toano Range in the extreme southeast corner of Elko County, 15 miles north of Ibapah, Utah. (See Pl. I.) The most eastern prospects of the district are about 2 miles west of the main freight road from Wendover, on the Western Pacific Railway, to Ibapah, and are close to the State line. Most of the prospects in the district are in a low, narrow east-west pass through the ridge and are within an area about 2 miles long by a mile wide. Beginning at the east, the properties on the north side of the pass are the Knowlton, Red Cloud, Big Chief, Martha Washington, and Ajax. The Salt Lake group lies on the south side of the pass opposite the Big Chief. The Regent mine is about 2 miles west of the Martha Washington, across the head of a broad, shallow valley which drains north. A 15-foot well in the bottom of Ferber Wash, about three-fourths of a mile east of the Knowlton property, furnishes all the water used, but, without question, water could also be obtained from shallow wells in Deep Creek Canyon opposite the mouth of Ferber Wash, where there is a small seep at which stock waters.

Geology.—The low pass is cut in a small stock of quartz monzonite that weathers more rapidly than the surrounding limestones. The unmetamorphosed limestones are blue-gray and are distinctly bedded in 1 to 2 foot layers. No fossils were found in them, but their lithology and their position on the eastern flank of the Toano Range suggest that they are of Carboniferous age. In general the beds strike east of north and dip southeast, but the structure near the mines is complicated and was not studied in detail.

The intrusive rock shows a rather wide range of texture, varying from a fairly coarse porphyritic quartz monzonite to a fine equigranular and somewhat gneissic rock that approaches granite. The normal porphyritic quartz monzonite has phenocrysts of andesine, quartz, and orthoclase (named in the order of decreasing abundance) in a fine granular intergrowth of orthoclase, quartz, and andesine. Biotite and hornblende occur both as phenocrystic minerals and in the groundmass. In some specimens 50 per cent of the rock consists of phenocrysts, but the proportions vary from place to place. The usual edge phase of this rock is finer grained and is more porphyritic than that from the center of the stock. At the Big Chief mine the rock is a fine equigranular aggregate of quartz orthoclase, andesine, and brown biotite. Gneissic banding has been developed to a slight extent near the north edge of the stock, but all gradations between the normal phase of quartz monzonite porphyry and the fine equigranular rock were noted.

The Regent mine is near the south contact of a small body of pyroxene-andesine rock (gabbro) that contains a small amount of interstitial quartz. In general it is a light-gray rock of fine granular texture, but in a few places it is of porphyritic texture. When examined in thin section under the microscope, it is seen to consist of gray-green augite and andesine with some orthoclase and a small amount of quartz. A dike of similar rock was noted about 50 feet west of the Martha Washington ore zone, cutting white crystalline limestone. This gabbro is possibly a differentiate of the quartz monzonite magma. though its relations to that rock are not definitely known.

Ore deposits.—The ore deposits of the Ferber district were discovered about 1880 by the Ferber brothers, who did a little work on the Big Chief vein and what is now the Salt Lake group. They did not hold their claims very long, and the properties were relocated about 1890. At present the largest property, consisting principally of the Martha Washington and Salt Lake groups, is owned by W. M. Bradley and associates, of Salt Lake. There are 15 patented claims in the district and a large number of locations. So far as could be learned the production from the district is quite small, consisting of less than 100 carloads of oxidized copper and lead ores which carry a little silver and gold.

The ore deposits, with the exception of the Regent and Big Chief, are typical contact-metamorphic deposits in crystalline limestone adjacent to the small stock of quartz monzonite. The contact alteration has resulted in the formation of lenslike masses of epidote, yellowish-brown garnet, calcite, amphibole, and quartz. The bodies of lime silicate are usually near the intrusive rocks, and are arranged in belts parallel to the contacts, which appear to dip away from the stock on all sides. The contact-metamorphic deposits are chiefly valuable for copper, which is found in the form of chrysocolla and copper pitch ore. Copper carbonate minerals are present in all amounts, but do not form a large part of the ore. In some of the deeper workings chalcopyrite was noted, and chalcocite was found in the workings of the Salt Lake group at a depth of 40 feet. It is probably also present in the other mines, though it was not seen by the writer. The Big Chief is a siliceous vein cutting quartz monzonite about 100 feet from its north contact. The surface ores are limonitic and contain lead carbonate and a minor amount of oxidized copper minerals.

The Regent ore body is a shear zone that strikes east and dips south in a basic phase of the quartz monzonite. The siliceous ore carries principally oxidized lead minerals and a small amount of copper carbonate at the surface. At a depth of 100 feet there is a streak of massive galena on the hanging wall.

The properties.—The Sidong claims, located by G. W. and S. A. Knowlton in January, 1908, lie on both sides of Ferber Wash, about 2 miles west of Deep Creek canyon. The principal development is a 40-foot whim shaft with 50 feet of drifting at the bottom. A drift trending S. 70° E., 40 feet in length, has evidently been driven to undercut an older 30-foot incline shaft sunk on the ore zone about 75 feet southeast of the whim. The ore body forms along an open fissure that strikes N. 20° E. and dips 80° W. Copper silicate and black copper pitch ore with some black oxide are the most abundant copper ores, but some malachite and azurite are present. The whim shaft is sunk on the contact of altered limestone and a small body of medium-grained quartz monzonite. Amphibole, epidote, and skeleton crystals of pink garnet are found in the zone of intense alteration, which is irregular but not wide. In the east drift the intrusive contact is peculiar, showing irregular fragments of altered limestone in the fine-grained edge phase of the igneous rock through a belt 4 feet wide. Postintrusive movement along the fissures has brecciated this rock and permitted the entrance of the solutions which deposited the copper minerals. In replacing the rock the mineralized solutions show a marked preference for the limestone, replacing it with chrysocolla and some copper carbonate. The igneous rock is less extensively mineralized.

Several other belts of copper-bearing lime silicate rock appear in open cuts on adjoining claims, both north and south of the canyon.

The Red Cloud workings, on a small flat south of the road in Ferber Wash, about a mile west of the Sidong claims, consist of several shallow pits sunk in red iron-stained siliceous croppings. The limonitic ore body occurs in a zone of brecciated white crystalline limestone on the north side of a vertical fissure that trends N. 60° E. and above a fault that strikes N. 45° W. and dips 40° SW. A little copper stain is seen here and there in the red ore and a few small patches of yellow lead carbonate were noted.

The Big Chief vein, the original discovery in the district, is about three-fourths of a mile west of the Red Cloud workings. The vein is 2 to 8 feet wide and stands well above the level flat on the summit of the pass. It strikes east and is nearly vertical, cutting a fine-grained, somewhat gneissic quartz monzonite. The vein is parallel to and about 100 feet south of the north contact of the igneous stock. It consists of white and dark-gray sugary quartz that on the surface is stained with iron and manganese oxides. The quartz has been brecciated and the crevices filled with limonite and with some copper and lead carbonates which are said to carry about $14 a ton in silver and lead. The vein is opened by several pits and by three shafts, the deepest of which goes down about 75 feet. None of the shafts could be entered. No sulphides were noted on the dumps.

The Martha Washington ground is at the west end of the stock about half a mile west of the summit of the pass. The deepest shaft on this ground, north of the road, is 100 feet deep, with a 50-foot drift west at the bottom. The collar of the shaft is in white crystalline limestone, which shows some lime silicate minerals and is about 15 feet west of a 20-foot belt of gossan that consists of epidote, garnet, and quartz. The zone of lime silicates, which dips about 60° W., crosses the shaft 40 feet below the collar, and is cut along its east side by the drift on the 100-foot level. A few residual masses of chalcopyrite and bornite, surrounded by copper pitch ore, are found at the 100-foot level. Both the croppings and the 40-foot level show some copper carbonate and silicate. Most of the ore shipped from this claim, said to average 12 per cent copper, has been taken from a 50-foot incline about 400 feet south of the road, in the ore zone. At this place the mineralization seems to be stronger than it is farther north, and the belt of lime silicate rock with abundant limonite and copper carbonates and silicates is 6 to 10 feet wide. It strikes N. 10° W. and dips about 50° W.

The Ajax workings on what appears to be the southward continuation of the Martha Washington ore zone, consist of a shaft 150 feet deep and of about 400 feet of drifting on the 75-foot and 150-foot levels.   Copper carbonate ores with lime silicate minerals and quartz have been found in a north-south belt that dips west at medium angles.

The Salt Lake group of eight patented claims and five locations is on the south side of the pass, opposite the Big Chief.   The principal work on the Covellite claim is a 100-foot shaft, but there are a number of short tunnels, inclines, and shafts at other places along the main ore zone.   The ore bodies occur in an east-west belt that dips south at low angles, in white crystalline limestone about 200 feet south of the quartz monzonite contact.   Garnet, epidote, amphibole, and calcite are abundant contact-metamorphic products, and the copper ores are found in lens-shaped masses with the lime silicate minerals.   At the surface copper carbonates and silicates are the chief metallic minerals.   Copper pitch ore is also present.   At a depth of 40 feet some chalcocite was noted in the oxidized copper ore, and it is said that native copper was found at this depth.   A few grains of chalcopyrite were noted in the ore on the dumps, but the development is not deep enough to have reached unaltered sulphides.

The Regent mine, belonging to Leffler Palmer, of Ibapah, is west of the head of a north-draining open valley that lies between Toano Range and Ferber Mountain about $2\frac{1}{2}$ miles west of the summit of Ferber Pass.   The main development is a 150-foot vertical shaft on the south side of a low ridge composed of gabbro or pyroxene diorite.   The collar of the shaft is about 20 feet south of the vein, which cuts across the shaft at a depth of 70 feet, and is 30 feet south of the shaft at the 300-foot level.   It is estimated that the vein will be 70 feet south of the shaft at the 150-foot level, which had not reached the vein at the time of visit.   At the surface a zone of yellow and red-stained crushed gabbro 20 feet in width, exposed in the open cut, strikes N. 62° W. and dips 70° SW.   It is said that the ore from the cut has a value of $15 a ton in lead and silver. Some of the lead is in the form of powdery cerusite, but plumbo-jarosite was also seen in the ore.   At the 100-foot level there is a streak of partly oxidized galena on the hanging wall of the ore zone. The 3 feet of ore next the hanging wall is reported to be of better grade than the remainder of the vein.

TECOMA DISTRICT.

LOCATION AND ACCESSIBILITY.

The Tecoma district (No. 11, fig. 1, p. 18) is in the northeastern part of the Goose Creek Hills, in northeastern Elko County.   The district is about 10 miles north of Tecoma, a town on the Southern Pacific.   The mines are on the west side of the eastern ridge of the hills, just west of the Utah-Nevada State line. (See Pl. I.) This portion of the Goose Creek Hills has a low rounded topography and is not much higher than the Tecoma Valley, which extends westward to Toano Pass at an elevation of approximately 4,600 feet. The Jackson mine, the most developed property in the district, is 11 miles north of Tecoma on the north side of a low east-west ridge which extends into the valley from the southeastern peak of the Goose Creek Hills. The Dunham claims are half a mile southeast of the Jackson on the south side of the same ridge; the Irish Boy group is about three-fourths mile east of the Dunham; and the Queen of the West shaft is about 2 miles east-southeast of the Jackson mine at the edge of the Tecoma Valley.

GEOLOGY.

The mines are strung along a low anticline whose axis strikes about N. 40° W. On the northeast limb of this anticline the beds dip 12°–20° NE. and on the southwest side 20°–35° SW. In the vicinity of the Queen of the West mine the axis of the anticline seems to have swung to nearly north-south and only the eastern limb is exposed. The lower beds are gray-blue limestones of Devonian age, containing fossils, among which Edwin Kirk, of the United States Geological Survey, has determined bryozoan (?) types and *Cyathophyllum* sp. The bryozoan, he states, is a peculiar type which ranges widely through the Devonian of the Western States. These limestones, in the vicinity of the Jackson mine on the northeast limb of the anticline, are about 500 feet thick and are overlain by 50 to 100 feet of dark-gray quartzite and underlain by a rather thin bed of white quartzite, below which there are shaly limestones. On the southwest limb of the anticline, in the vicinity of the Dunham claims, thin-bedded sandy lime shales and limestones of light-gray and drab color carry fossils, among which G. H. Girty has determined *Fusulina* aff. *F. elongata* and fragments of crinoids. These beds are regarded as probably the equivalent of the White Pine shale, of the Eureka district. The anticline is somewhat broken by north-northwest fractures, along which the ore bodies are localized.

In the vicinity of the Dunham and Queen of the West properties there are some poorly exposed small dikes and sheets of a very much altered porphyritic rock which carries large phenocrysts of quartz and soft altered feldspar in a fine-grained groundmass. A thin section of this altered granite porphyry shows the groundmass to consist of quartz, sericitized orthoclase, some bleached biotite, and what appear to have been hornblende crystals, now altered to iron oxide.

On the flat southwest of the mines there are a few low buttes composed of pinkish-gray rhyolite in which flow structure is well developed, so that in places the rocks have a shaly cleavage.

ORE DEPOSITS.

HISTORY AND PRODUCTION.

The Jackson mine was discovered in 1906 and has been a small producer ever since. The latest location, the Irish Boy group, was made in January, 1912. The following table shows the production from the Tecoma district as published by the United States Geological Survey in Mineral Resources of the United States.

Production of the Tecoma district, Elko County, 1907–1912.

Year.	Gold.	Silver.	Copper.	Lead.	Total value.
	Fine ounces.	*Fine ounces.*	*Pounds.*	*Pounds.*	
1907..	3.48	946	....	59,000	$3,830
1908..	9.53	5,628	1,803	294,310	15,779
1909..	8.17	5,488	85	314,861	16,563
1910..	....	788	....	56,229	2,899
1911..	.48	349	127	20,968	1,154
1912..	.05	8	....	5,352	249
	21.71	13,207	2,015	750,720	40,474

CHARACTER OF DEPOSITS

The ore deposits of the Tecoma district are small irregular replacements in limestones of Devonian or Mississippian age and are localized along fissures which strike N. 15°–34° W. The relation between the ores and the obscure granite porphyry intrusions is not shown. The ores so far developed at all the properties consist of argentiferous cerusite, which is more or less ocherous in the northwestern part of the district and which contains some zinc carbonate in the mines to the east.

THE PROPERTIES.

Jackson mine.—The Jackson mine on the north side of a low ridge about 11 miles north of Tecoma is in bluish-gray northeast-dipping Devonian limestones. The ore bodies occur along a N. 15° W. fracture which dips 35°–50° W. They are replacements but do not appear to favor any particular bed of the 500 feet of limestone exposed on the ridge. The upper workings are tunnels whose total length is 500 feet in the ore zone, which is about 20 feet wide. In these workings the upper limit of the ore is a bed of white quartzite. The main development is a 115-foot inclined shaft with about 35 feet of drifting at the lower level and many irregular raises and stopes between that level and the surface. The small irregular ore bodies occur throughout a zone 25 feet wide.

The ores are ocherous lead carbonates, of which some from the lower level carry a small quantity of galena and anglesite. The sorted ore is said to average 25 per cent lead and 10 to 12 ounces of silver a ton, though some shipments carrying 44 per cent lead and 20 ounces of silver have been made. The property is reported to have produced $50,000 worth of ore since its discovery in 1906.

Durham claims.—The four claims of the Durham group are on the south side of the ridge half a mile south-southeast of the Jackson mine. The main development work is a 60-foot vertical shaft which starts on a fracture that strikes N. 30° E. and dips 60° SE. A 15-foot drift on the 30-foot level intersects the ledge. Next the hanging wall of the fracture there is 4 inches of clay, below which the irregular and rather small ore bodies make in lime shales that dip 45°–50° SW. The ore is a soft canary-yellow lead carbonate and is said to average 40 per cent lead, 32 ounces silver, and $2.50 in gold a ton.

Irish Boy group.—The Irish Boy group, located January, 1912, by Carl Peterson and Dave Douglas of Grouse Creek, Utah, is 1¼ miles east-southeast of the Jackson mine. A 60-foot vertical shaft, equipped with a small gasoline hoist, constitutes the main work, but is apparently in barren ground. From the surface to a depth of 10 feet the shaft is in granite porphyry, below which there are thin-bedded cherry blue limestones. There was apparently no contact metamorphism along this sheet of porphyry.

Queen of the West claims.—The Queen of the West group of five claims, the farthest southeast of any property in the district, is apparently on the eastern limb of the anticline, where it has swung into a north-south trend. The cherty bluish-gray limestones dip about 45° E. and are cut by an irregular fracture which strikes about N. 20° W. and dips 75° ENE. The ore occurs near this fracture as an irregular replacement of the limestones. It is a dark cellular cerusite ore with a minor amount of limonite and a small quantity of smithsonite. The high-grade sorted ore is said to carry about 80 ounces of silver a ton and in some shipments as high as 18 per cent of zinc, though the usual zinc content is about 2 per cent. The main development is an inclined shaft 120 feet deep on the main fracture.

LANDER COUNTY.

LOCATION AND TRANSPORTATION.

Lander County, in the central part of Nevada, has excellent transportation facilities. The main line of the Southern Pacific crosses its northern end, and the narrow-gage Nevada Central Railroad follows Reese River through its center. At both Battle Mountain and Austin supplies of all kinds can be obtained. The roads in the county are good and numerous lines of stage operate from both terminals of the railroad, giving easy access to practically all parts of the county.

LOCATION AND ACCESSIBILITY.

The Ravenswood district, in south-central Lander County (No. 15, fig. 1, p. 18), is about 20 miles northwest of Austin, the county seat. Most of the prospects are on the summit of the Shoshone Range, a few miles southwest of Ravenswood Peak, though one property is on the eastern side of the range near the head of Reese River canyon. This prospect is about 1½ miles northwest of Vaughn siding on the Nevada Central Railroad. Ravenswood camp is 7 miles west of Silver Creek siding, which is 70 miles south of Battle Mountain, the junction with the Southern Pacific main line.

TOPOGRAPHY.

Along the west side of Reese River valley, extending from Silver Creek to Vaughn siding, a broad bench lies 150 to 200 feet above the river flat. West of this bench Ravenswood Peak rises 2,000 feet in a little less than 3 miles. Its eastern face is rough and is cut by deep, steep gorges, but to the west it falls off in a gradual slope to Antelope Valley. This peak marks the northern end of the southern portion of the Shoshone Range, whose northern portion lies beyond Reese River, which crosses it through a deep, narrow canyon about 6 miles to the north. Southward from the peak a ridge approximately 6,500 feet high extends for several miles.

GEOLOGY.

Classes of rocks.—The higher parts of the range are composed of dark quartzites, shales, and arenaceous limestones, though on some of the ridges north of Ravenswood Peak lighter-colored bluish-gray limestones are present. At the Rast mine, 1½ miles northwest of Vaughn siding, the sediments are intruded by dikes of quartz monzonite porphyry. A large part of the lower slopes of the range are underlain by rhyolitic and andesitic lavas, remnants of which also lie on the summits of the ridges, as though the whole range had once been buried under the volcanic rocks.

Sedimentary rocks.—The older sedimentary rocks exposed in the Sell Canyon road about 5 miles northwest of Silver Creek siding are dark brownish-drab micaceous quartzites and dark-greenish and drab micaceous shales, which strike N. 20° W. and dip 30° W. Among a few fossils collected from a bed of shale at this locality Edwin Kirk, of the United States Geological Survey, recognized the Cambrian forms *Hyolithes* sp. and indeterminable fragments of trilobites. These quartzites and shales appear to be at least 500 feet thick, and are overlain by 120 feet of lighter-colored quartzite, 500 feet of yellowish-drab, red, and bluff thin-bedded calcareous shales, 300 feet of gray iron-stained quartzites with some shales, and at least 1,700 feet of dark-gray crystalline limestones in medium thin beds. Calcite veinlets cut these limestones in all directions, and black chert is abundantly distributed throughout the formation. The exact age of these formations is not known, but it is thought that they are all of Cambrian age and may possibly be correlated with the lower part of the section at Eureka, Nev., described by Hague.[1]

In the vicinity of the Ravenswood camp the main structural feature is the persistent north strike and west dip of the sediments. The dip is 45° W. at the head of Sell Canyon, but west of the camp diminishes to about 15°. A secondary structure at the camp has thrown the westward-dipping beds into a shallow synclinal fold so that a short distance south of the camp the beds dip 35° to 45° NW. A large number of the properties are located in this syncline along northeast-trending fractures which have broken the beds.

At the Rast mine, near the head of Reese River canyon, the nearly black limestones and lime shales strike N. 10° E. and dip 45° W. On the summit of the ridge southwest of the mine the dark rocks are overlain by lighter-colored heavy-bedded limestones.

The benches along Reese River between the head of the canyon and Silver Creek are underlain by light-buff partly consolidated sands and silts, which were referred by the geologists of the Fortieth Parallel Survey to the Tertiary Truckee formation.[2] These beds are exposed only on the sides of the washes, where the rhyolite that covered them has been eroded.

Along the Reese River bottoms there are small areas of Quaternary silts which are extremely fertile. In this part of its course Reese River flows above ground, and most of this bottom land is irrigated and under cultivation.

Igneous rocks.—The limestones at the Rast mine have been intruded by a number of dikes of light-gray to white, rather coarse-grained, somewhat porphyritic quartz monzonite. A few of these dikes exposed in the crosscut tunnel have been more or less altered, and most of them contain some disseminated pyrite. A thin section of this rock examined microscopically shows it to consist of about equal amounts of feldspar and quartz and a very minor proportion of light-green biotite. Magnetite and apatite are rare accessory minerals. The feldspars, particularly the plagioclase, which seems to be near andesine, and the microperthite are sericitized, but the

[1] Hague, Arnold, Geology of the Eureka district, Nev.: U. S. Geol. Survey Mon. 20, pp. 34–41, 1892.

[2] Emmons, S. F., U. S. Geol. Expl. 40th Par. Rept., vol. 2, p. 639, 1877.

more abundant orthoclase has suffered only slight alteration. The normally dark-gray to black limestone is changed to a nearly white, coarsely crystalline rock in the vicinity of these dikes, but no lime silicate minerals were noted at any place.

So far as determined there are no intrusive rocks in the vicinity of Ravenswood camp, and from information obtained from the operators in the district it is thought that none are nearer than those near the Rast mine or those to the south-southeast, which were spoken of as granites.

Along the eastern face of the range there are thin flows and flow breccias, pinkish drab to red-drab in color, which so far as seen are typically developed rhyolites. Along the Sell Canyon road the material resembles a fine breccia of fragmental spherulitic rhyolite with a few quartz and biotite fragments. Near the Rast mine the rocks are thin flows of porphyritic rhyolite containing rather abundant small phenocrysts of quartz, orthoclase, and biotite in a partly glassy groundmass containing orthoclase quartz and biotite. Rhyolites are also present on the western slopes of the mountains.

The rhyolites are younger than the light-colored friable sandstones of the Truckee formation, which they cover on the eastern side of the range. They are in turn covered in some places to a considerable depth by deposits of gravel and wash from the mountains.

On the summit of the range at Ravenswood camp remnants of andesitic flows cap the higher parts of some of the ridges, notably the ridge between Ravenswood camp and the Queen shaft. At this place the reddish biotite andesite showing a distinct flow lamination is about 60 feet thick. At other places the remnants are 5 to 15 feet in thickness.

ORE DEPOSITS.

HISTORY.

Ravenswood, one of the early organized districts in the Reese River country, is at the northern terminus of the Shoshone Mountains. According to Browne,[1] silver was the principal metal sought in early days, and as the veins near Ravenswood Peak were rich only in copper, they were abandoned by the disappointed prospectors. According to White,[2] the veins that occur in a belt 10 miles long and 2 miles wide on the east side of the Shoshone Mountains in limestone and slate were discovered in 1863 by a party from Austin. The Shoshone ore deposit, located in February, 1870, is a northwest-

[1] Browne, J. R., Mineral resources of the States and Territories west of the Rocky Mountains for 1867, p. 413, 1868.

[2] White, A. F., Nevada State Mineralogist Third Biennial Rept., for 1869–70, p. 44 [1871].

southeast vein 6 feet wide that dips 45° NE. and carries more galena than copper. It seems probable that this property is now known as the A. J. C. No. 5 claim, upon which there is an old abandoned shaft from which a small shipment of lead ore is said to have been made by Mr. Clayton in 1895. The main properties were later controlled by Mr. Oddie and associates, who inaugurated a short-lived boom in 1906 and 1907. At the present time most of the claims are held in two groups, one by Mr. J. M. Pine, of Denver, Colo., and the other by an association of people from Ely, Nev. A few claims are held independently.

So far as can be learned there has been a very small production from the camp, and all of the ore was apparently mined previous to 1902, for the statistics of mineral production collected by the United States Geological Survey since that year show no output from Ravenswood.

ECONOMIC CONDITIONS AND DEVELOPMENT.

Transportation is not difficult in this district, as the railroad is less than 12 miles from the most remote workings and is within 3 to 4 miles of many of the prospects. There are fairly large piñon and juniper trees on the range, some of which grow near the camp. Water is abundant along the Reese River valley but in the mountains occurs only in a few small springs, that at Ravenswood camp, supplying about three or four barrels a day, being one of the larger.

The deepest development, the Queen shaft, is down about 240 feet. The Red Bird shaft, said to be 80 feet deep, was filled with dirt and water in the latter part of September, 1913. Most of the claims have a few 10-foot prospect holes on the croppings of the veins and lenses. The development of the district is really insufficient to give a basis for economic geologic study.

TYPE OF DEPOSIT.

In the vicinity of Ravenswood the ores occur in small lenslike quartz veins, which either parallel or cut the rather indistinct bedding of dark shales, quartzites, and limestones of Cambrian age. These veins, consisting of both light and dark quartz and occasionally barite, carry some chalcopyrite, antimonial galena, and tetrahedrite, all of which sulphides are said to carry silver and a little gold. The sulphides are seen at the surface at many places, but are somewhat oxidized at the greatest depths obtained in the prospects. The chalcopyrite usually alters to blue or green chrysocolla and more rarely to azurite. Galena alters to anglesite, cerusite, and occasionally to lead phosphate (pyromorphite), occurring as a yellowish-green coating on joints in the ore and country rock near the veins. The carbonates of copper and lead are only slightly developed. The quartz in many veins is barren or only slightly mineralized. The veins themselves are small, and none of them have been traced for any considerable distance on the surface. Many of the quartz bodies have the appearance of lenses rather than veins.

THE PROPERTIES.

Pine properties.—On the Portland claim, about one-fourth mile south-southeast of Ravenswood camp, there are a few shallow surface workings on a deposit of oxidized ore in a brecciated yellowish-brown shale lying above a flat strike fault. The shales strike N. 33° E. and dip 75° NW. In one cut the ore goes to a depth of 4 feet to the nearly horizontal fault plane, below which what appear to be similar shales show no mineralization. Between the fragments of the breccia, quartz, minor amounts of blue or green copper silicate, and some yellowish earthy lead phosphate have been deposited.

On the next claim south of the Portland a vein trending N. 60° W. and dipping 80° NE. is exposed by the surface workings. The country rocks, drab and yellow arenaceous shales, with some yellow limestones, strike N. 60° W. and dip 50° N. The narrow white quartz vein is stained with blotches of copper carbonates surrounding small particles of a gray-black copper sulphide which appears to contain antimony, but blow-pipe tests on this mineral were not entirely conclusive.

The Red Bird shaft, about 3½ miles south-southwest of the camp, is in the canyon bottom. The cloud-bursts of the summer of 1913 filled this shaft, which is said to have been 80 feet deep on the incline, with débris and had carried away a considerable portion of the dump. The ledge strikes N. 40° W. and dips 50° SW. parallel to the bedding of the slate and limestone country rock. The hanging-wall rock is a thin-bedded buff lime shale and the footwall a greenish shale. The ledge is about 4 feet wide and has more the form of a lens than a vein, though it seems to be traceable for about 500 feet along the surface. It consists of dark-gray smoky quartzite, in which a small amount of pyrite and chalcopyrite is irregularly distributed. On some of the joints there are films of copper silicate and carbonate. As seen in thin sections under the microscope this siliceous material is seen to be a quartzite in which the outlines of the original quartz grains are marked by bands of black inclusions.

About half a mile west-northwest of the Red Bird, on the southeast side of the canyon in whose head Ravenswood camp is located, a short crosscut tunnel and shallow shaft open up a 3-foot ledge of silicified black shale stained with copper silicates that is said to carry good-grade silver ore. This ledge strikes N. 15° E. and dips 75° W., cutting the formation, which in this place strikes N. 10° E. and dips 65° W.

On the A. J. C. No. 4 claim, about a mile southwest of the Queen shaft, a 25-foot incline develops a 2-foot lenslike body of quartz stained with copper silicate and lead carbonates. A little residual chalcopyrite was seen in the ore and a few specks of what appeared to be galena.

On the A. J. C. No. 5 claim, about three-eighths mile south of the camp, there is an old inaccessible inclined shaft, reported to be 80 feet deep, from which some ore is said to have been shipped. A few pieces of siliceous ore on the dump are stained with chrysocolla and carry galena, gray copper, and stibnite. On the surface the vein seems to strike N. 10° W., and to dip 70° E. It is said to have varied from 2 to 3 feet in width through the workings.

The Queen shaft, about one-fourth mile southwest of Ravenswood camp, on the southeast side of the canyon, is equipped with a small gasoline hoist. In the middle of September, 1913, it was approximately 240 feet deep. There is a 120-foot drift on the 154-foot level with a 75-foot crosscut, about 80 feet of drifting on the 70-foot level, and a drift tunnel about 120 feet long which intersects the shaft 32 feet below the collar. The dark limestones and shales in this ridge strike N. 45° E. and dip 20° NW.

On the surface the vein strikes N. 32° E. On the 70-foot level it strikes N. 35° E. and dips 85° SE. and is 6 feet wide, with fair walls, and consists of quartz, with some partly oxidized copper and lead sulphides. At the tunnel level the vein is 8 to 10 inches wide, with a smooth, well-marked footwall along which there has been some postmineral movement. The hanging wall of the vein is ill defined. The wall rocks on this level are calcareous shales and limestone.

The mineralization along this vein was spotty, there being considerable white and dark quartz filling without any sign of metallic minerals. At other places, irregularly scattered along the strike, small masses of sulphides occur. Chalcopyrite and silver-bearing antimonial galena seem to have been the chief primary minerals, though what appeared to be gray copper was noted in some of the ore on the dump. As a rule, the minerals are not intergrown and no age relations were determined. The galena alters to anglesite and minor amounts of cerusite, and thin films of the greenish-yellow lead phosphate are present on fractures in the ore. The chalcopyrite alters to bluish and greenish silicates, with very minor amounts of carbonate.

In the shaft, at a depth of 80 feet, a flat northeast-dipping fault cuts off the vein, and in none of the work below that level was ore seen. It was unfortunate that the shaft was in such condition that this break could not be studied in detail and a definite idea obtained of the direction or amount of movement along it; the movement, however, appears to have been rather small, as the rocks are little disturbed near the plane and there is not much gouge. In the crosscut on the 154-foot level there is a N. 10° W. fracture that dips 65° E.; along which there has been a normal movement which does not appear to have been great. This can not be the same as the flat fault seen in the shaft.

Ely properties.—A number of claims on the summit and eastern side of the range, controlled by a group of men from Ely, Nev., are locally known as the Giroux Ely group.

The Independence No. 2 claim, a mile east-southeast of Ravenswood camp, on the ridge south of the head of Sell Canyon, is developed by a few shallow workings on a 2 to 3 foot bedded vein in shales that strike N. 50° E. and dip 35° NW. The vein consists of white and dark quartz in which there are small amounts of copper silicates, galena, and chalcopyrite.

On the Independence No. 1 claim the lode strikes N. 45° W., stands vertical in some places, and dips 40° N. in other pits. It is exposed in four prospect pits over a length of 250 feet. It consists of two or three narrow quartz stringers that are subparallel, but branch and join along the strike. A little lead and copper stain is seen on much of the quartz, but so far as noted no sulphides have been found.

The Lead King ground lies south of the Independence claims and about 1½ miles southeast of Ravenswood camp. A 10-foot shaft reveals a 4-foot lens of white quartz, somewhat stained with limonite and copper silicate, that strikes N. 30° E. and dips 85° E., cutting dark-gray and brownish limestones which have been much disturbed. At the bottom of the shaft there is some antimonial galena and a little gray copper.

On the Shamrock No. 1 claim a prospect pit has exposed four narrow white quartz veins irregularly spaced in a 4-foot vertical zone of brecciated gray-brown crystalline limestone which strikes N. 35° W. These stringers continue with their average width in buff and greenish shales interbedded with the limestone. A small amount of copper silicate is present in this quartz. At another shallow pit a N. 20° E. vertical quartz vein carries, beside copper silicate, some partly oxidized chalcopyrite.

In six pits on the Shamrock No. 2 a vertical quartz vein is exposed, which strikes N. 20° E., cutting the interbedded shales and limestones. It consists of white and gray quartz and is not strongly mineralized. At some places it reaches a minimum of 5 feet but averages about 2 feet in width. In the narrow portions chalcopyrite is seen in the quartz, altering to copper pitch ore and copper silicates.

On the Emmet claim, three-fourths mile south of the Shamrock, a north-trending vein which dips 45° W. cuts the shales at an acute angle. The vein is 2 feet wide and consists of white quartz, which has been crushed and partly recemented. Copper silicates, limonite, and a little residual chalcopyrite were seen in the ore on the dump, which is said to carry gold, silver, lead, and copper.

The Wild Horse claim is a mile south-southeast of Ravenswood. A small lenslike replacement consisting of somewhat iron-stained quartz, barite, and calcite, carrying a little copper carbonate, lies in lime shales above a bed of quartzite.

Lombardy group.—The Lombardy group, commonly called the Rast property, is on the southwest side of Reese River, near the head of the canyon and about 1½ miles northwest of Vaughn siding. The vein, which outcrops near the summit of a north-south ridge, is opened by two shallow inclines, from which a little ore is reported to have been shipped. The vein strikes N. 10° E. and dips about 50° W., nearly parallel to the bedding of dark limestones. It is 8 to 16 inches wide, with fair walls. The ore is an antimonial silver-bearing galena, with possibly a little gray copper. Anglesite and some cerusite are the alteration products. A parallel vein, opened by a few shallow pits farther up the hill, carries chalcopyrite and sphalerite, as well as the minerals seen in the first vein. A crosscut tunnel, now 400 feet in length, has been started from the base of the ridge east of the veins, which have not been cut as yet. In this tunnel the limestone is crystalline and much sheared along a series of fractures which strike N. 10°–20° E. and dip 60°–70° W. There has also been some movement along practically vertical east-west fractures. A light-colored, rather coarse grained quartz monzonite has intruded the sediments but has caused little or no contact metamorphism. Pyrite in small cubes is abundant in some dikes and in the limestone adjacent to several of the intrusive masses. As seen in the tunnel the dikes are irregular, very narrow, and have been sheared and crushed since their injection.

Other prospects.—The Last Chance, 3½ miles south of Ravenswood, is a 4-foot quartz vein that cuts the bedding of yellow and green shales which strike N. 20° E. and dip 40° W. Chalcopyrite and antimonial galena and their oxidation products are sparingly distributed through the vein.

On the Warrior claim a ledge of what appears to be a dark quartzite strikes N. 40° W. and dips west and is more or less stained with copper silicate.

LINCOLN COUNTY.

MINERAL DEVELOPMENT.

Lincoln County, established in 1866 from part of Nye County, included all of southeastern Nevada. In 1906 the southern half was set apart as Clark County, with Las Vegas as the seat. Pioche, one of the old Nevada mining camps, was flourishing in the sixties, and has been the seat of Lincoln County since 1869. For many years Milford, Utah, was the nearest railroad point. In 1873 the narrow-gage railroad was built from Pioche to the smelter at Bullionville, now known as Panaca. In the seventies and early eighties 110 stamps were working on Pioche ore, and at least two smelters have been built to treat the product of the Meadow Valley district. In 1907 Pioche was connected by rail with Caliente, on the San Pedro, Los Angeles & Salt Lake Railroad.

The mines in the Atlanta, Bristol, and Patterson districts of north-eastern Lincoln County were visited during the later part of October. They are grouped about a valley which lies east of the south end of the Schell Creek Range and north of Pioche. (See Pl. I.) On all the maps of Nevada this valley is named Duck Valley, pre-sumably from Duck Lakes, which are south of Geyser post office at the north end of the valley, but according to the recorder of Lincoln County [1] the northern end of this valley south to an east-west line through Bristol Pass should be called Lake Valley, and south of that line, where the valley narrows, it should be called Patterson Wash. According to a recent decision of the United States Geographic Board, this valley will in the future be known as Lake Valley.

ATLANTA DISTRICT.

LOCATION AND ACCESSIBILITY.

The Atlanta district (No. 16, fig. 1, p. 18) is about 40 miles north-east of Pioche, its shipping and supply point, in a parklike basin on the northeastern flank of the Fortification Range. This range, really a spur of the Schell Creek Range, separates Lake Valley on the south-west from Spring Valley on the northeast. (See Pl. I.)

The mines are north and west of a prominent, though relatively low mountain, which rises slightly above the rolling hills on the northeastern side of this low divide. (See fig. 8.)

GEOLOGY.

Spurr believed these mountains, which he called the Cedar or Clover Mountains, to be entirely composed of lavas [2] similar to the rocks examined by him along Meadow Valley Canyon,[3] where he observed the following sequence of events: .

1. Deposition of the Paleozoic series of quartzites and limestones.

2. Elevation of this series to a land mass and the erosion of the rocks to produce a system of mountains and valleys. This was attended by little or no folding.

---

[1] Letter dated June 17, 1914.

[2] Spurr, J. E., Descriptive geology of Nevada south of the fortieth parallel and adjacent portions of California: U. S. Geol. Survey Bull. 208, pp. 36–37, 1903.

[3] Idem, pp. 139–148.

3. Pouring out of great masses of white biotite rhyolite (early Tertiary).

4. The formation of a series of water-laid rhyolitic sandstones and tuffs, interbedded with thin sheets of rhyolite and rhyolite breccias. This whole series is roughly estimated at 4,000 feet thick and at the top contains relatively more tuffs, while at the bottom there are relatively more lavas. Several slight unconformities and many slight erosion gaps occur in the series.

5. Folding to a considerable degree of the whole crust.

6. Explosive eruptions of pyroxene andesites and latites of moderate extent.

7. The formation of a series of water-laid brown volcanic tuffs or sandstones and breccias, with interbedded quartz-bearing volcanics, chiefly dacites and reddish rhyolites. The sandstones were relatively thick at the bottom of the

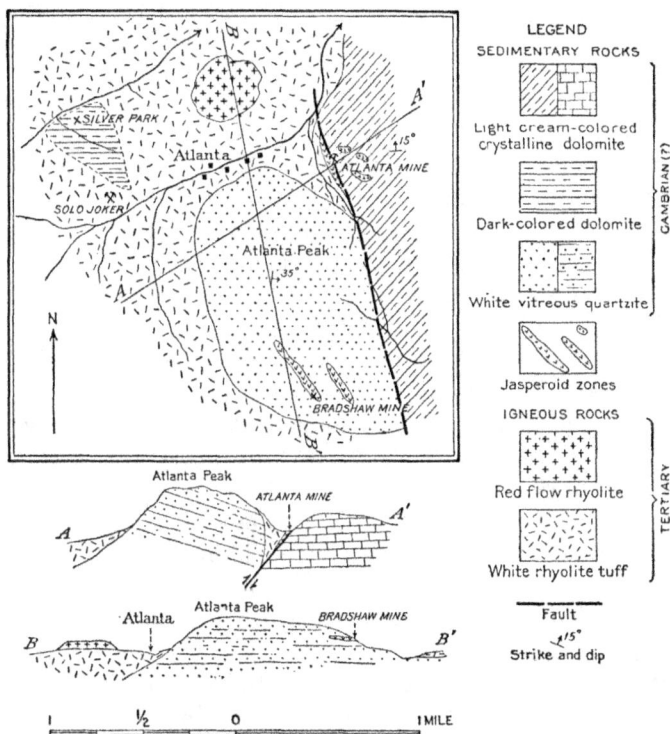

FIGURE 8.—Sketch map and sections of the Atlanta district, Lincoln County, Nev.

series, the volcanics at the top. The entire thickness of the series is estimated at 3,500 feet. There are some petty erosion intervals.

8. General folding, comparatively gentle.

9. Deposition of at least 2,000 feet of brown or red conglomerates and soft sandstones, which are accompanied by very few volcanic flows and so are distinct from the preceding formations. They have remained nearly horizontal and are probably, in large part at least, lake beds. They have been referred to the Pliocene.

10. Drainage of the Pliocene lake, erosion, and slight local folding in the Pleistocene.

11. Outpouring of thin sheets of rhyolite, tordrillite, and pyroxene-olivine basalt.

12. The formation of a small amount of high-stream gravels.

13. Cutting down of the canyon bed to its present position.

The thickness of the basal rhyolite is not known. A very roughly estimated section of the overlying formations is as follows:

Section in Meadow Valley Canyon.

	Feet.
Rhyolite tuff series	4,000
Andesite	600
Red lava and sandstone	3,500
Pliocene sandstones and conglomerates	2,500
	10,600

The observations of the writer confirmed Spurr's conclusion in so far as the western and southern part of these mountains is concerned, but it seems that a considerable area in the northeastern part of the range is underlain by sedimentary rocks, which are buried beneath the volcanic rocks. At Atlanta the mountain south of the town is composed of a fine-grained vitreous white quartzite, in thick beds which dip east-northeast at medium to low angles. (See fig. 8.) The ridge east of the town is composed entirely of light-gray crystalline limestones in beds 2 to 4 feet thick, which strike east and west and dip 15°–30° N. These two formations are separated by a fault which strikes N. 30° W. and dips 60° SW. (See fig. 8.) North of Atlanta Peak dark limestones apparently overlie the quartzite, though this relation is not certain, as a detritus-filled canyon separates the exposures.

No fossils were found in the sedimentary rocks, so their age is not known, though it is believed possibly to be Cambrian or Ordovician, as white quartzites at least 1,000 feet thick, overlain by 4,000 to 5,000 feet of dark and light limestones of Cambrian age,[1] form much of the south end of the Snake Range, which lies northeast of Atlanta across Spring Valley.

A light-gray to white biotite rhyolite tuff overlies both the quartzite and limestone and has been involved in the faulting. (See fig. 8.) About 200 feet of the tuffs are exposed. They are overlain by dark-red flow rhyolites, which form a large number of the buttes north and west of Atlanta. It is the belief of the writer that the rhyolite tuffs and flows are probably the equivalents of No. 4 of Spurr's section. (See p. 115.)

ORE DEPOSITS.

History.—Silver was mined in the late sixties and early seventies in the Silver Park district, which is now known as Atlanta. These

---

[1] Howell, E. E., U. S. Geog. Surveys W. 100th Mer. Rept., vol. 3, p. 241, 1875. Weeks, F. B., Geology and mineral resources of the Osceola mining district, White Pine County, Nev.: U. S. Geol. Survey Bull. 340, pp. 119–122, 1908.

mines were evidently known to the geologists of the Wheeler Survey, for the position of the Silver Park district is accurately shown on a map of eastern Nevada which accompanies the progress report for 1869.[1] So far as known, no descriptions of the mines in this vicinity have been published; it is impossible to find any account of the early production of silver from this camp; and no production has been reported to the United States Geological Survey since 1902. It seems highly probable, however, that the Atlanta mine will be a producer in the near future, and it is known that lessees on the Silver Park property made some shipments of high-grade silver ore in 1913.

Occurrence of the ores.—At the Atlanta, Bradshaw, and Solo Joker mines the ores occur in zones of cemented breccia along zones of faulting. At the Bradshaw the rich silver minerals occur in fractures in a dense, very hard, red jasper. The ore zones at the Atlanta and Solo Joker are more open breccias, consisting of fragments of limestone, rhyolite tuff, quartzite, chert, and jasper, partly cemented by quartz, and containing much limonite and some pyrolusite. Some of the ore from these properties has the cellular platy structure commonly developed in veins of the late Tertiary period of mineralization. In this kind of ore the quartz is seen to be replacing barite. The fact that these ores occur in faults involving the rhyolite tuff, and that fragments of the tuff are included in the silicified breccias, is clear evidence that they were formed after the extrusion of those lavas which Spurr assigns to early Tertiary time. At all the mines of the district silver and gold are the valuable constituents of the ores, and so far as known the base metals play a very minor part in the mineralization. At all of the properties silver is an important constituent of the ore, though at the Atlanta mine gold is the most important. Very minor amounts of copper and lead carbonates are present in most of the ore, but limonite and manganese oxide are present in all of it.

The Silver Park deposit is different from the others of the district in that the ores occur in limestone in small quartz veins that have the appearance of replacement veins along open watercourses. So far as known silver is the only valuable metallic constituent of the ores, though copper and lead carbonates are present in most of the thoroughly oxidized material mined from this deposit.

THE MINES.

Atlanta mine.—The Atlanta group of 11 patented claims and 5 locations was acquired by the Atlanta Consolidated Gold Mining & Milling Co. in 1906. The group covers the north and east sides of Atlanta Peak, and the principal development work is on the

---

[1] Wheeler, G. M., U. S. Geog. Surveys W. 100th Mer. Prelim. Rept., 1869.

fault (see fig. 8) east of that mountain. The vertical shaft, equipped with a 50-horsepower steam hoist, was 225 feet deep in October, 1913, with about 350 feet of crosscuts and drifts on the 100-foot level and much less work on the 200-foot level. The No. 2 shaft, about 500 feet to the south, is 75 feet deep, and some shallow workings one-fourth mile south of the hoist shaft are on the continuation of the ore zone.

The main ore zone lies above or west of the fault, which separates limestone on the east from quartzite and rhyolite tuff on the west. This fault strikes approximately N. 30° W. and dips 42° WSW. in the main shaft and in the workings one-fourth mile south. At the latter place the fault clearly separates limestone on the footwall (east side) from quartzite on the west. Here it is an open crevice, between walls that in places are 2 to 4 feet apart. There was apparently little mineralization at this locality.

The main shaft starts in rhyolite tuff. It cuts the ore zone about 90 feet below the surface and passes out of it near the 200-foot level. At the 100-foot level the drift eastward shows the ore breccia to be 150 feet wide and to rest upon a crystalline limestone footwall, and drifts north and south show it to be at least 100 feet long. The ore breccia consists of fragments of rhyolite tuff (replaced by silica), chert, and limestone, partly cemented by white quartz. Thin sections of the altered tuff, examined under the microscope, show it to be completely silicified into a fine inequigranular aggregate of quartz. Some of the quartz is clear, but most of it is crowded with minute black inclusions, in many places arranged in plumelike forms. When examined under high power, some of the larger inclusions are seen to be red and translucent and are thought to be hematite, but the character of the great majority can not be determined.

In some of the ore angular cells surrounded by narrow bands of quartz have replaced plates of barite. These resemble the quartz-adularia veins formed in late Tertiary time. A careful examination of several thin sections of this type of ore failed to show any adularia such as is commonly present in this type.

The fine-grained free gold of this ore is largely carried in the limonitic material that partly cements the breccia. This consists predominantly of limonite, pyrolusite, and minute white quartz crystals. Some gold is said to occur in the hard silicified rhyolite tuff. On the limestone hill east and north of the Atlanta shaft, large croppings of red jasper along a north-northeast course intersect the main fault at an acute angle. It is said that all these croppings carry some gold, and that the main ore body, 75 feet wide, averages $4.25 in gold and silver a ton.

Specimens of rhyolite tuff and ore, said to have been found at a depth of 145 feet in the main shaft, are coated on joint planes with canary-yellow, finely crystalline carnotite that contains both uranium and vanadium. It was apparently deposited in joints formed after the gold mineralization of the ore.

Bradshaw mine.—The old Bradshaw mine is on the south end of Atlanta Mountain about a mile south of the Atlanta mine. (See fig. 8.) It is developed by a large open cut and an incline shaft about 75 feet long with 50 feet of drifting near the surface.

The workings are on the western and larger of two zones of red jasperoid which strike N. 45° W. and dip 50° NE. The western zone, 500 feet long and 150 feet wide, is a firm hard silicified breccia of quartzite, limestone, and jasper cemented by red jasper. Narrow open crevices and stringers, which strike N. 60° W. and dip 40° NE., cut the jasper and are said to carry the rich oxidized silver ores. No recognizable silver minerals were seen in the small amount of second-grade ore on the dump; though a few purplish-gray iridescent stains may be very finely divided particles of silver haloids. Limonite in minutely botryoidal form was seen in the crevices.

Silver Park mine.—The Silver Park mine, known also as the Jesse Knight property, is a little over a mile west of the Atlanta shaft. In October, 1913, it was leased to Devlin and Hoskins, who are reported to have shipped during the summer a carload of ore that carried 100 ounces of silver to a ton. This property is said to have been worked in the late seventies. The shaft is in limestones south of a shallow gulch. Porphyritic rhyolite tuffs overlie the limestone north of the gulch, the contact striking N. 70° E. and dipping 40° N. The old shaft is said to be nearly 100 feet deep but was obstructed 50 feet below the surface. The workings consist of small, very irregular, inclined drifts and stopes which follow open watercourses. The main channel along which mineralization has occurred strikes east, but other channels trend north. In these watercourses large crystals of " dog-tooth " calcite have been deposited and considerable manganese oxide and limonite filled into the drusy openings. Between the walls and the calcite and also along what appear to be bedding planes are narrow quartz veinlets which constitute the ore. The silver minerals are all of the oxidized type, and so far as known no original sulphides have been found. The yellow stains on the ore are argentiferous lead carbonate. Aggregates of minute green and blue crystals are silver-bearing copper carbonates, and thin purplish-green stains seem to be a mixture of copper and lead carbonate. The richest ore is coated with irregular dirty brown films of soft waxy horn silver. Some of these scales of horn silver cover as much as a square inch, and are one-sixteenth inch thick. A thin section of a piece of rich ore shows horn silver scattered through the quartz as well as in the joints. A few specks of a blue-gray metallic mineral, which is semitranslucent and is decidedly red in reflected light, proves to be ruby silver.

Solo Joker claim.—Two claims about a mile west of the Atlanta mine, one of which is known as the Solo Joker, are owned by the Bank of Pioche. These claims are half a mile south of the Silver Park mine. In this vicinity the rhyolite tuff is about 20 to 50 feet thick, covering the dark limestones. At the main workings a small area of limestone is exposed on top of a ridge. The ore body is a silicified breccia of rhyolite tuff, limestone, and jasper 10 to 15 feet wide, and is in the form of an angle, one leg of which strikes N. 60° E. and the other N. 35° W. Some quartz seen on the dump looks like the platy cellular quartz-adularia vein material of the late Tertiary deposits. When examined in thin sections the quartz is seen to be replacing barite and to be crowded with minute black and red inclusions, many of which have plumelike forms. Limonite, hematite, and pyrolusite occur as coatings in the drusy cavities. No metallic minerals were seen in the ore, which is said, however, to carry gold and silver.

PATTERSON DISTRICT.

LOCATION AND ACCESSIBILITY.

The Patterson district (No. 18, fig. 1, p. 18) covers the south end of the Schell Creek Mountains, in Patterson Pass. South of this low pass the same uplift is known as the Ely Range, whose southward extension, in the vicinity of Pioche, is known as the Highland Range. The eastern entrance to Patterson Pass is about 45 miles north of Pioche, and the main road from that town to Ely, in Lake Valley, passes within 4 miles of the old mining properties in the summit of the pass. (See Pl. I.) Several claims have been located in Swartz Canyon, about 3 miles north of Patterson Pass.

TOPOGRAPHY.

The southern part of the Schell Creek Range is a very narrow north-south ridge, higher than any of the near-by ranges except the Snake Range. Patterson Peak, about 10,000 feet in elevation, is 6 miles north of the pass, whose elevation is approximately 7,250 feet. Lake Valley, 4 miles east of the summit of the range, is about 6,500 feet above sea level. Cave Valley, west of the mountains, probably somewhat higher than Lake Valley, is long and narrow, looking more like a canyon than the usual desert valleys of Nevada. From both valleys the mountains rise in steep slopes and in many places in cliffs of considerable height.

GEOLOGY.

SEDIMENTARY ROCKS.

The geology of the Schell Creek Range in the vicinity of Patterson Pass was described by Howell as follows: [1]

From Patterson, which is at the southern termination of the range, northward for 25 to 30 miles, the rocks * * * appear to dip toward the east at high angles. * * * At Patterson a heavy bed of quartzite is exposed. * * * A few miles farther north this is covered with conformable bluish-gray limestone; limestone was also seen to the west of the quartzite at Patterson, apparently forming with it a faulted anticline. * * * This limestone agrees lithologically with that of the Snake Range to the east and of the Highland Range to the south and is underlain with a similar bed of quartzite.

Spurr,[2] in reviewing the geology of southern Nevada, says:

Inasmuch as at least the southern portion of the Highland Range consists of Cambrian limestones, which were classed by Mr. Howell as Carboniferous, * * * and since the same Cambrian series occurs in the Snake Range, it seems likely that these rocks near Patterson may also be Cambrian.

As seen by the writer the quartzites and overlying limestones of this portion of the Schell Creek Range appeared to be very similar to the series on the east side of the range in the vicinity of Aurum and Schellbourne. Spurr[3] obtained Cambrian fossils from the shales lying between the quartzite and limestone near Schellbourne.

At the western side of Lake Valley limestones, poorly exposed and largely covered by gravels and silts, seem to conformably overlie the quartzites which form all of the eastern flank of the mountains in this vicinity. In the lower part of Swartz Canyon the limestones overlie 400 feet of light-colored quartzite, which in turn overlies a great thickness of dark reddish-brown quartzite. The quartzites continue to the divide at the head of Swartz Canyon, but north of the saddle they are overlain by 100 feet of dark shales and mica schists that are covered both north and south by gray crystalline limestones which weather in light colors. The lower limestone beds are somewhat shaly, but at Patterson Peak the limestones are in thick massive beds. Patterson Pass is cut in massive dark crystalline limestones in beds, some of which are 50 feet thick. Spurr[4] states that Cambrian fossils were found by Weeks in the limestones above the quartzites near Patterson. Some thin beds of dark shale are interbedded with these limestones.

---

[1] Howell, E. E., U. S. Geog. and Geol. Surveys W. 100th Mer. Rept., vol. 3, p. 242, 1875.

[2] Spurr, J. E., Descriptive geology of Nevada south of the fortieth parallel and adjacent portions of California: U. S. Geol. Survey Bull. 208, p. 40, 1903.

[3] Idem, p. 39.

[4] Idem, p. 20, footnote.

So far as seen this part of the Schell Creek Range is devoid of igneous rocks with the exception of a narrow granite porphyry dike which cuts across the divide at the head of Swartz Canyon. This dike, 50 feet wide, strikes N. 45° W. and is nearly vertical. It weathers readily and is inconspicuous. Where exposed in some shallow pits the rock is iron stained, fine grained, and is speckled with biotite plates one-eighth inch in maximum diameter and with a number of pinkish phenocrysts that were at first thought to be orthoclase, but that, when examined in thin section, proved to be calcite, which presumably had replaced the original feldspars. The mica is remarkably fresh in view of the apparent altered condition of the hand specimen. The largest part of the rock is composed of quartz of two generations and of orthoclase. The texture is peculiar in that grains of clear quartz are surrounded by quartz carrying abundant inclusion prisms of orthoclase. The halos of the matrix quartz extinguish at the same time as those of the original quartz, yet have a somewhat radial appearance which is accentuated by lines of minute black inclusions.

There is very little contact metamorphism along the dike, but about one-fourth mile northeast of the summit lime shales, lying above the true shale and below the massive limestones, have been altered to masses of quartz, calcite, diopside, fluorite, sphalerite, and pyrite.

STRUCTURE.

In Swartz Canyon the sedimentary beds all strike north and south and dip east. At the mouth of the canyon the dip is 36°, but near the summit it is about 15°. Just east of the divide a strong north-south fault which seems to dip east at a steep angle has broken the strata. West of this fault the beds appear to have been lowered relatively to those east of the break. The quartzites and overlying limestones near the summit of the range dip east at low angles, but about half-way down the west slope east dips as high as 60° were recorded. (See fig. 9.)

Halfway between Patterson Pass and Swartz Canyon a nearly vertical fault that strikes N. 50° W. interrupts the general structure of the range. South of this fault the limestones have been brought against the quartzites, which occur north of the break. Near the fault the bedding strikes N. 45° E. and dips 50° SE., but farther south in the pass the strike is N. 18° E. and the dip 15°–20° ESE.

ORE DEPOSITS.

HISTORY.

Rich silver ores were found in Patterson Pass about 1869,[1] though the deposits may have been known to the Indians before their discovery by the whites.[2] Soon after the discovery over 200 claims were located in the hills north of the pass, where all of the deposits had been found. The ores were rich oxidized material carrying $212 to $520 in silver and $22 in gold a ton. White reported that lead, antimony, copper, and iron were present in the ore. Apparently the activity of this camp was short lived, and there does not seem to have

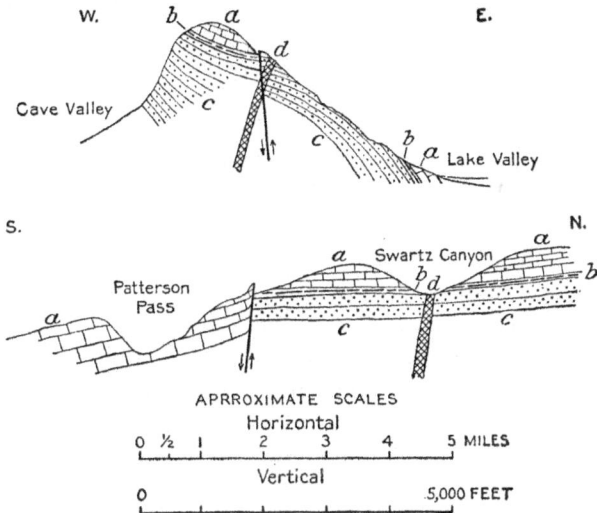

FIGURE 9.—Diagrammatic sections of the south end of the Schell Creek Range near Patterson Pass, Lincoln County, Nev. East-west section along Swartz Canyon; north-south section at summit of range. a, Gray crystalline limestone, weathers light colored; b, dark shales and schists; c, dark-colored quartzite; d, intrusive granite.

been any deep mining on any of the claims. There are no records of production from the district, either in the early days or in the last decade.

OCCURRENCE OF THE ORES.

In the limestone hills north of the pass some small replacement veins of white quartz and calcite strike N. 10°–20° E. and are nearly vertical. The walls are frozen in most places, though in one shaft postmineral movement has formed a thin selvage approximately on the east wall of a narrow quartz vein. The quartz and calcite are

---

[1] Raymond, R. W., Statistics of mines and mining in the States and Territories west of the Rocky Mountains for 1869, pp. 178–179, 1870. Wheeler, G. M., Preliminary report upon a reconnaissance through southern and southeastern Nevada in 1869, p. 12, 1875.

[2] White, A. F., Nevada State Mineralogist Third Biennial Rept., for 1869–70, p. 93 [1871].

well crystallized and metallic minerals are not abundant. Occasionally thin yellow films of argentiferous lead carbonate are seen on quartz crystals, and more rarely small pockets of greenish-blue silver-bearing copper carbonates are present.

Near the summit, on the west side of the mountains about 2 miles north of the pass, a group of claims belonging to E. L. Robertson cover a fault which strikes N. 50° W. A shallow incline northeast of the fault is driven on an iron-stained body of crushed shale immediately above quartzite. This lenslike zone has a maximum width of 8 feet and averages about 4 feet. Near the center of the zone 10 inches to 2 feet of crushed rock coated with copper carbonates constitutes the ore. It is cut off at the south by the fault. One hundred feet south of these workings, still northeast of the fault, a 50-foot tunnel runs north along the contact of the shale and quartzite. At this place strike faulting has formed a heavy gouge on the hanging wall, which dips 15° E. Below the gouge the crushed shale contains a small amount of copper carbonates as thin patchy coating.

On the summit of the range at Swartz Canyon Mr. Jacob Swartz is developing a group of claims by a crosscut tunnel, which runs N. 25° W. for 300 feet through a dark, slightly calcareous quartzite. About 100 feet above the tunnel an iron-stained cropping of altered lime shale lies at the base of the limestones. In this zone the shale is altered to a heavy dark-green rock consisting of quartz, calcite, diopside, and fluorite, which carries some sphalerite and pyrite. At the level of the tunnel the quartzites are unmetamorphosed, but there are thin films of sphalerite and pyrite in the joints which cut the formation. In one rather strong east-west joint with steep north dip a few scales of molybdenum were seen.

BRISTOL DISTRICT.

LOCATION AND ACCESSIBILITY.

The Bristol district (No. 17, fig. 1, p. 18) is near the center of the Ely Range, about 15 miles north-northwest of Pioche. Bristol Pass crosses the range 2 miles north of the camp. Most of the mines are on the west side of the mountains, though one of the largest properties, the Day (Jack Rabbit) mine, is on the northeast side. A narrow-gage railroad connects this mine with Pioche.

TOPOGRAPHY.

Most of the mines are in a cove on the south side of an east-west spur of the range (see fig. 10) at elevations ranging from 7,200 to 8,300 feet above sea level. The summit of this ridge is nearly 8,500 feet in elevation. Both north and south of the summit the steep slopes are broken by cliffs and the topography is rough. South and west of Bristol low foothills extend to Bristol Valley, whence the water supply for the camp is obtained from wells. North of the summit the slopes are particularly steep, there being a drop of nearly 2,000 feet in a little over a mile between the summit and the Day (Jack Rabbit) mine, which is near the upper edge of the long alluvial slope on the southwest side of Lake Valley. The ridge is notched by short, steep, dry canyons with almost vertical sides.

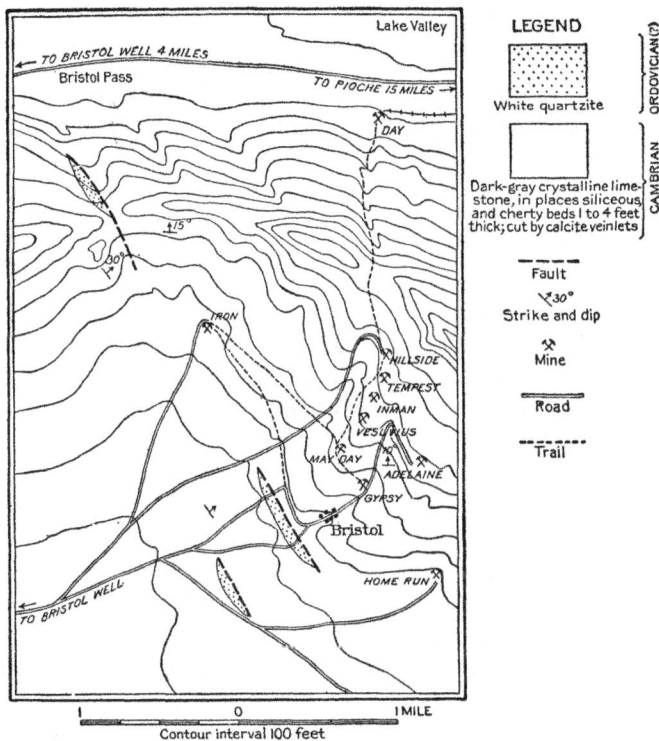

FIGURE 10.—Sketch map of the Bristol district, Lincoln County, Nev.

GEOLOGY.

This part of the Ely Range is composed of limestone of a prevailing dark-gray to black. Some of the beds are crystalline, but many are very dense, hard, and siliceous and in places look like argillites or fine-grained quartzites. Dark shales are interbedded with the limestones, particularly near the base of the section exposed south and east of Bristol camp. Most of these limestones are in beds 1 to 4 feet thick and are distinctly bedded. It is believed that this limestone is the equivalent of the Hamburg limestone of the Eureka section. Practically all of the beds strike east and dip from 5°–10° N. Minor local variations occur in the general strike and dip, but nowhere are the sediments strongly folded and nowhere do they exhibit steep dips. West of the main fault along the west base of the mountains the limestones and overlying quartzites dip east-northeast. Near the fault, dips as high as 30° were recorded, but farther west low dips again prevail.

So far as known all the ore deposits occur in the dark limestones which are referred to the Cambrian.

Walcott[1] described the Cambrian section in the west side of the Highland range halfway between Bennetts well and Stampede Gap as follows:

Summary of Cambrian section of the Highland Range west of Pioche.

Bottom not exposed.	Feet.
Quartzite _____	350
Limestone and shales, argillaceous and arenaceous_____	1, 450
Massive limestones_____	3, 000
	4, 800

He further states that the limestone member of this section may be 5,500 feet thick in the mountains north of his measured section. The white quartzite on the west side of the range (see fig. 10) was correlated with the Ordovician Eureka quartzite of the section at Eureka, Nev., by Walcott, who adds: " No strata overlying the upper quartzite were observed in the Highland range between Bennetts Springs and 2 miles north of Bristol."

Along the west side of the productive district there is a fault whose course is roughly indicated by outcrops of the brecciated fine-grained white quartzite. (See fig. 10.) It seems possible that there may be two subparallel faults southwest of Bristol. On the ridge northwest of the camp but one fault was noted, and it appeared to stand vertical. Along the western side of the fault the white quartzite has been dropped against the underlying Cambrian limestones. The magnitude of the faulting was not worked out in this reconnaissance, but it would appear that the relative displacement must be at least 1,500 feet. Spurr[2] says of this fault:

Between the Silurian foothills and the Cambrian rocks of the main range there appears to be a great break, bringing about juxtaposition of strata which in their normal stratigraphic succession are separated by nearly 2 miles of intervening sediments.

In the main body of the limestones there are many minor fissures. One set strikes east-northeast and dips south at medium angles; the other set is nearly parallel to the main fault west of the range. The

[1] Walcott, C. D., Second contribution to the studies on the Cambrian faunas of North America : U. S. Geol. Survey Bull. 30, pp. 33–36, 1886.

[2] Spurr, J. E., Descriptive geology of Nevada south of the fortieth parallel and adjacent portions of California : U. S. Geol. Survey Bull. 208, p. 46, 1903.

north-northwest fissures usually dip east at steep angles, though some dip west.

So far as known, no intrusive igneous rocks occur in this part of the range. One body of iron-stained material, which occurs along the main fault zone a mile south of the Home Run mine, was locally thought to be a porphyry, but proved, on examination, to be a fault breccia composed of rather coarse fragments of limestone and quartzite in a fine calcareous sandy matrix.

MINERAL DEPOSITS.

HISTORY AND PRODUCTION.

According to Whitehill,[1] the Bristol district was organized April 10, 1871, by Hardy, Hyatt, and Hall. The ore deposits of this region were known in the late sixties, as most of this region had been prospected shortly after the discovery of Pioche, which took place in 1863. A smelter was built at Bristol Well in the late seventies, and later, a 5-stamp mill at the same place.[2]

In 1877 the Hillside Co. was organized to take over the properties of Mr. Steele, which included the Hillside mine and the well and smelter. At that time the incline on the Hillside fissure was 200 feet deep,[3] and the vein was reported to be 5 to 8 feet wide, carrying ore which averaged $100 a ton in silver and lead.

The Census report for 1880[4] says that at the Hillside and Day mines galena and cerusite ores in limestone were being mined.

For several years the Hillside and Bristol companies operated the Hillside and May Day deposits on the west side of the range. The Day (Jack Rabbit) mine in Lake Valley was operated independently. These properties and several others near Bristol were consolidated in 1911 by the Day-Bristol Consolidated Mining Co., which, in 1913, was operating the May Day and Gypsy mines and had leased the Inman, Tempest, and Hillside properties. The Day (Jack Rabbit) was idle, and it was reported that the lower workings were below the ore zone. This company at present holds 23 patented claims and 4 locations.

The production of the mines in the vicinity of Bristol previous to 1904 could not be learned, though undoubtedly many thousand dollars worth of silver, lead, and copper were recovered during the period from 1869 to 1904.

---

[1] Whitehill, H. R., Nevada State Mineralogist Fourth Biennial Rept., for 1871-72, pp. 111-112 [1873].

[2] Abbott, J. W., Pioche, Nev.: Min. and Sci. Press, vol. 95, pp. 176-179, 1907.

[3] Whitehill, H. R., Nevada State Mineralogist Seventh Biennial Rept., for 1877-78, p. 77, 1879.

[4] Emmons, S. F., and Baker, G. F., Statistics of technology of the precious metals: Tenth Census, vol. 13, p. 36, 1885.

Production of mines in the Bristol district, Lincoln County, Nev., 1904–1912.

	Gold.	Silver.	Copper.	Lead.	Total value.
	Fine ounces.	*Fine ounces.*	*Pounds.*	*Pounds.*	
1904...................................	48.39	26,580	....	100,000	$20,000
1906...................................	43.05	37,649	442,520	284,000	127,709
1907...................................	11.32	15,624	103,845	13,552	32,115
1908...................................	13.79	143,420	273,386	294,976	124,774
1909...................................	13.54	12,867	162,815	86,233	31,824
1910...................................	....	999	6,581	....	1,375
1911...................................	10.57	143,598	12,086	28,063	79,099
1912...................................	4.35	7,134	33,581	11,889	10,809
	145.01	387,871	1,034,814	818,713	427,705

OCCURRENCE OF THE ORES.

All of the ores mined in the Bristol district occur as replacements of limestone of Cambrian age. As a rule the replacement of the limestone was most intense at the junctions of two sets of more or less open fissures. One set of fissures strikes N. 45°–60° E. and dips 45°–50° SE.; the other set strikes N. 10° E. to N. 10° W. and as a rule dips steeply east. The larger ore bodies of the May Day, Gypsy, and Hillside occur above or southeast of fissures of the east-northeast series of fractures and east of the north-south fissures and are usually lenticular in shape. The ore deposits along the north-south series of fissures are more commonly thin tabular bodies that somewhat resemble veins but that are clearly replacement deposits.

CHARACTER OF THE ORES.

The ores so far mined in the Bristol district are almost entirely oxidized. Lead carbonate ores containing some zinc carbonate and carrying more or less copper and silver are found in most of the mines. Copper ores consisting of chrysocolla, copper pitch ore, and lesser amounts of the copper carbonates, are found in other deposits. In the Gypsy and Hillside mines both classes of ore occur in the same stopes. Near the hanging wall of the shoots lead carbonate ores practically free from copper minerals are found; but on the footwall copper ores predominate, in some places to the exclusion of lead minerals. Limonite and the black sooty manganese oxide pyrolusite are abundant constituents of the ores in the copper stopes and are present in smaller quantities in most of the lead ores. Silver is carried in both the lead and copper ores, but it seems to be the general experience that the copper carries more silver than the lead.

Residual nodules of galena surrounded by thin envelopes of anglesite are found in some of the lead carbonate ores. A specimen of ore said to have come from the lower levels of the Hillside mine consists of chalcocite mixed with a little galena. The chalcocite is coated with crusts of malachite, which also cuts the sulphide in veinlets. Galena and chalcocite were the only sulphides seen by the writer in the ores of this district.

The bodies of oxidized copper and lead ores always occur above planes which would interfere with the downward circulation of waters. They usually occur above or south of the northward-dipping, eastward-trending fissures, and on the hanging-wall side of the northward-trending fissures. Most of the few bedded deposits exposed in various mines occur above shaly or dense siliceous beds which have prevented the downward movement of the waters. The deposits so far opened are with little question thought to have been formed by downward-moving waters and are probably the result of a long-continued process. The source of the metals is not clear, as no deposits of primary sulphides have been opened in the district. It may be that deposits of the original sulphides will be found at greater depths on the fissures, but from all that is now known it seems as reasonable to suppose that the oxidized ores were derived from sulphide deposits that have been entirely removed.

Since the ore bodies now being mined were formed by downward-moving waters, it is of some interest to know to what depth they may be expected to continue. It is to be presumed that primary sulphide ores, if any, will not be present in large quantities above the ground-water level. As there are no springs in these mountains, it must be admitted that the water table is lower than the bottoms of any of the canyons. That it is probably considerably deeper than most of the canyons is evidenced by the dryness of the lower workings at the Jack Rabbit mine, which are at least 1,800 feet lower than the lowest workings of the Gypsy.

The possible continuation of the ores with depth is another serious question which can be answered only by development; yet it may be mentioned in this connection that below the 900-foot level in the vertical shaft at the Day mine there seems to have been no considerable mineralization, and that even on the 600-foot level long stretches of the fissure are barren of ore bodies, to judge from the stope maps.

Features of the operations.—Most of the better known and most developed mines of the Bristol district are controlled by the Day-Bristol Consolidated Mining Co. The group of 23 patented claims and 4 locations extends from Bristol across the range northward beyond the Day (Jack Rabbit) mine. This company also controls the wells 5 miles west of Bristol camp, which are the sole water supply of the camp. The more important mines controlled by this company, named from south to north, are the Gypsy, May Day, Vesuvius, Inman, Tempest, Hillside, Iron, and Day (Jack Rabbit). In 1913 all the ore mined at these properties either by lessees or the company was hauled by team to the terminus of the narrow-gage railroad at the Day mine at a cost of $4 to $6 a ton. In October, 1913, the plans for an aerial tram over the range from the Day to the Gypsy mine were completed, and it was expected that this tram would be completed early in 1914. When this tram is in operation it should reduce the freight charges by about $4 a ton.

Gypsy and May Day mines.—The Gypsy and May Day mines, controlled by the Day-Bristol Co., are on the ridge a short distance north of Bristol camp. (See fig. 10.) The lower shaft, the Gypsy, is vertical, and was about 500 feet deep in October, 1913, with drifts at 100, 135, 150, 300, 350, 400, and 500 feet. The May Day incline is sunk 41° SSE., about on the dip of the fissure. It was 500 feet long with levels 100, 200, 250, 300, 350, and 500 feet on the incline. The bottom level of the May Day incline connects with the 300-foot level of the Gypsy shaft. There is a 22-horsepower gasoline hoist and a 1-drill compression at the May Day and a 50-horsepower hoist and 5-drill compression at the Gypsy.

On this ridge the dark siliceous limestones strike east and dip 10° N. They do not appear to have suffered any folding and show little displacement along the fissures that were the channels of the mineralizing solutions.

The Gypsy ore body occurs along a fissure that strikes N. 6°–10° W. and dips 80° E. and the May Day along one that strikes N. 60° E. and dips 43° S. On the 300-foot level the May Day fissure cuts across the Gypsy fissure 250 feet north of the vertical shaft. (See fig. 10.) At the 350-foot level the intersection is 150 feet north of the shaft; near the 400-foot level what seems to be the May Day fissure is exposed in the Gypsy shaft; and on the 500-foot level a fissure with a strike similar to the May Day is exposed in the workings 100 feet south of the Gypsy shaft.

The principal mineralization of these fractures appears to be near their intersection. Movement along the May Day which seems to have taken place prior to the formation of the carbonate ore bodies has cut off the Gypsy vein. The probable continuation of the Gypsy fissure is seen on the 300-foot level of the May Day about 50 feet east of the place where the drift from the Gypsy shaft intersects the May Day workings. The northward continuation of the Gypsy is represented by two tight fissures about 4 feet apart that strike N. 43° E. and dip 85° E. (see fig. 11), along which there has been only slight mineralization.

The Gypsy fissure is a more or less open watercourse. In places it opens into caves lined with calcite crystals or, more rarely, with stalagmites of calcium carbonate. Brownish red limonite and some soft black manganese oxides coat the drusy surfaces. Along this fissure the limestone has been replaced by the carbonate ore minerals, but the extent of replacement varies widely, and the ore bodies are very irregular in detail. The larger ore bodies of the Gypsy seem to occur near the intersection with the May Day fissure, and as a consequence pitch 40° to 45° S. In the upper levels of the Gypsy shaft, which are farther from the junction than the lower levels, the ore bodies were not so large as those stoped between the 250 and 400 foot levels. On the 100-foot level the copper stopes south of the shaft are 4 to 6 feet wide and lie on the footwall of the fissure below a narrow lime horse. They are in medium-grade ore traversed by a central 18-inch streak of high-grade copper carbonate. Above the horse there is a streak of lead carbonate ore of varying width. At the 150-foot level the ore body was 14 feet wide, the lower 10 feet carrying copper carbonate and the upper 4 feet lead carbonate. From the 250-foot level

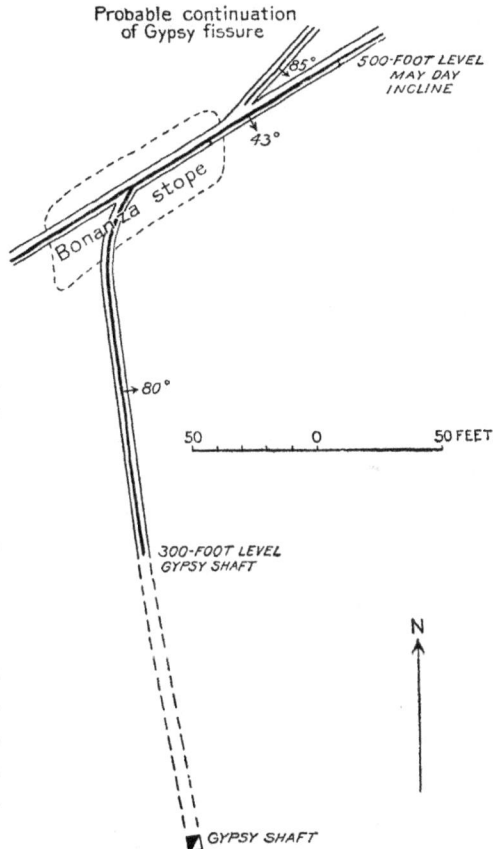

FIGURE 11.—Sketch map to show the probable continuation of the Gypsy fissure north of the May Day vein, Bristol district, Lincoln County, Nev.

to the greatest depth attained at the time of visit there was a nearly continuous shoot of ore 20 to 40 feet wide and averaging about 150 feet long which pitched 45° S. on the vein. This ore body occurs above the May Day–Gypsy intersection and has been mined in the Perry and Green stopes and the Loyd winze. In the stopes in this ore shoot nearly pure lead carbonate ore was found near the hanging wall and copper ores next the footwall.

Lead carbonate said to carry some silver was the only lead mineral noted in the mine. The chief copper ores are copper pitch, malachite, chrysocolla, and a copper-bearing iron oxide. The copper pitch ore varies from a black substance which resembles pitch to an earthy reddish or greenish compound containing some copper. The black variety appears to consist of copper and manganese with almost no iron. By far the greater part of the copper ore is a copper-bearing limonite in which the copper seems to be in the form of red copper oxide, probably in large part in the small masses of the carbonate minerals irregularly scattered throughout the limonite.

Along the May Day fissure the mineralization was apparently not so intense as along the Gypsy, though at the intersection of the two there were large bodies of ore. Some normal movement, which displaced the Gypsy fissure and produced a rather heavy clay gouge on the footwall, seems to have taken place. The incline shaft is in the fissure for 300 feet, at which depth is cut the footwall, and at the 500-foot level is 200 feet north of the fracture. At this level the vein is tight in much of the drift, but opens in places to a width of 2 feet. In the more open portions of the fissure some high-grade copper pitch and carbonate ore was deposited in shoots which seem to pitch 30° E. on the vein. One ore shoot of copper-bearing iron oxides on this fissure extends from the surface to a depth of 200 feet. On the 100-foot level it is 300 feet long and about 60 feet wide.

A streak of rich copper pitch ore on the hanging wall of this body has been mined. The May Day ore bodies apparently carry very little lead, but contain some zinc at the 300-foot level. Considerable silver is reported in the limonitic copper ore. The Bonanza stope is in a body of good-grade limonite copper ore which occurred at the junction of the May Day and Gypsy fissures but was longest on the strike of the former fissure. It extended from the 300 to the 500 foot levels and was about 120 feet long. A streak of rich black copper oxide followed the hanging wall of this ore body.

Vesuvius mine.—The Vesuvius ore bodies, which lie on the summit of the ridge about one-fourth mile north of the May Day incline and 200 feet above it, are bedded replacements of limestones which dip 5° N. The replacement occurs south of a N. 60° E. fissure, which dips 45° S. and which is exposed in the back of the lower flat room-like stope. The largest body of ore was west of a N. 16° W. vertical fracture on the east side of the stopes. Two bedded deposits, about 50 feet apart, are exposed in the workings. The lower consists of lead and zinc carbonate that clearly replaces dark limestone. The ore body was apparently 2 to 3 feet thick, had a fairly regular floor but an irregular roof and lateral limits and was about 50 by 100 feet in largest horizontal dimensions. The upper bed, 18 to 24 inches thick, carries much more copper carbonate than the lower bed but contains both lead and zinc carbonates. It makes above and south of the fissure that trends N. 60° E. and west of the vertical north-south break. Both ore bodies occur in nearly pure limestone above beds of very dense siliceous shaly limestone.

Inman mine.—The Inman mine, about 1,000 feet north of the Vesuvius mine, is in a body of copper carbonate ore, said to carry both zinc and lead, which occurs in several fissures. (See fig. 12.) These fissures cut a dense hard dark-gray siliceous limestone which has not been replaced by ore to any considerable extent. Practically all the ore occurs between the walls of the fissures and seems to have resulted from the replacement of brecciated rock. The fissure which strikes N. 80° E., dips 60° S., and is followed by a 100-foot tunnel, is 2 to 4 feet wide. It is filled with a limonite ore carrying copper, lead, and some zinc in the form of carbonates. In the 20-foot winze near the face of the tunnel some small bedded replacements were noted. The Inman vein, which strikes N. 25° W. and dips 70° E., has been underhand stoped to a depth of 50 feet. It is 10 inches to 2 feet wide, with well-defined walls of dense black limestones which have not been replaced by ore minerals. The vein is filled largely with dark-red to black copper pitch ore and with some chrysocolla. About 40 feet west of the Inman a nearly parallel fissure 1 to 3 inches wide, filled with copper-stained gouge in the wider portions but practically barren in the narrower, is exposed in a 15-foot open cut. The shaft 80 feet west of the Inman vein is on a fissure which strikes N. 46° W. and is nearly vertical. It is from 10 to 18 inches wide and carries oxidized copper ores.

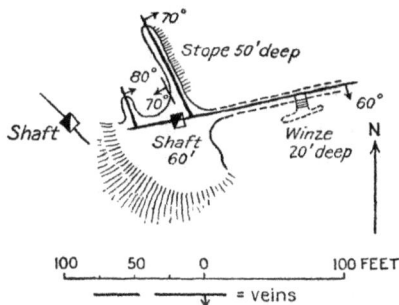

FIGURE 12.—Inman mine, Bristol district, Lincoln County, Nev.

Tempest mine.—The Tempest mine, at an elevation of 8,100 feet, is about a mile north of Bristol. (See fig. 10.) The 350-foot incline shaft has short drifts at the 60, 150, and 200 foot levels and about 350 feet of drift on the 300-foot level. The workings are on a fissure which strikes N. 80° E., dips 80° S., and is filled with 10 inches to 2 feet of crushed limestones and gouge. The irregular bodies of oxidized copper ores occur in the hanging wall. At the 150-foot level a body of ocherous ore 4 to 6 feet wide carries small amounts of copper carbonate. On the 300-foot level east of the shaft a fairly continuous streak of copper pitch ore 2 to 4 inches wide lies above the gouge. This ore is said to carry some silver. The shaft has been equipped with a small steam hoist by the lessees, Nesbit and Bolling.

Iron mine.—The Iron mine, owned by the Day-Bristol Consolidated Co., but under lease to J. A. Nesbit, is about a mile northwest
of Bristol camp. (See fig. 10.) The fissure is opened by three shafts,
the southernmost of which was being worked in October, 1913. This
shaft was 380 feet deep, with four levels about 90 feet apart. The
longest drifts on the 200-foot level total 165 feet, mostly south of the
shaft. Above this level the ore has been stoped to the surface. The
ores make in irregular lenses 6 feet in maximum width in either the
foot or hanging wall of the fissure, which strikes N. 12°–15° E. and
dips 48° E. The ore is soft iron-stained lead carbonate, and the
lower-grade ore contains much unreplaced limestone. A considerable
quantity of high-grade sorted ore seen on the dump is hard massive
cerusite. In this class of ore rounded residual kernels of galena are
surrounded by layers of anglesite and cerusite. A very minor amount
of copper carbonate is present in some ore, and thin films of silica
have been deposited in openings and crevices in the ore. Two grades
of ore are sorted at the mine, neither of which is said to carry much
silver. Freight charges to Pioche from this mine are $4 a ton, and
about 55 tons a month are shipped.

Hillside mine.—The Hillside mine is the most northern of the
mines on the west side of the range and is not far below the summit,
at a barometric elevation of 8,225 feet. In the fall of 1913 it was
leased by the Hillside Leasing Co., which at the time of visit was
installing a gasoline hoist and was cleaning out the old workings.
This mine, one of the early producers of the camp, was most extensively worked between 1870 and 1885. The incline is 900 feet in
depth, with drifts about 100 feet apart. The second, seventh, and
eighth levels were open for some distance, but the remainder of the
old workings could not be entered for more than a few feet. The
incline is on the fissure to the fifth level, but below that it has
been sunk in the hanging wall.

The Hillside ore body occurs along a strong fissure which strikes
N. 60°–63° E. and dips on an average 40° SE. In the upper
150 feet the dip is about 75° S. The increase of dip near the surface may be due to a subparallel vertical fissure which occurs 35
feet below the main vein at the 200-foot level and intersects the
vein at the surface. Both of these veins were stoped to the 200-foot
level.

It is said that the main ore shoot was 120 feet long and that the
shaft was near the center of the shoot. Between the second and
fifth levels the ore zone is said to have averaged about 16 feet wide.
At a few places near the shaft between these levels pillars of ore
showed two and some of them three slips cutting the ore about parallel to the footwall. At the seventh level, about 400 feet below

the collar of the incline, the ore body averages about 14 feet wide. It is divided into two distinct classes of ore. Next the hanging wall a fairly continuous narrow belt of lead carbonate is separated from the lower ore body by a horse of limestone 2 to 6 feet thick. The lower ore body varies from 6 to 10 feet thick and consists of a mixture of limonite, copper carbonates, and copper pitch ore. This ore body continues to the eighth level and has been stoped for some distance southwest of the shaft. On the eighth level, 300 feet southwest of the shaft, the Hillside fissure is cut off by a vertical fracture that strikes N. 15° E., along which there has been some mineralization. The ore body at this intersection has been stoped and it was impossible to determine the nature of the fault movement. In the copper ore body on this level some masses of sulphide have been found. A specimen of this material, said to assay 54.05 per cent copper, 72.5 ounces silver, 4 per cent lead, and 0.04 ounce gold a ton, proves to be a mixture of chalcocite, galena, and malachite. The carbonate occurs as veinlets cutting and coating the sulphide, which on first inspection seemed to be entirely chalcocite.

Day (Jack Rabbit) mine.—The Day (Jack Rabbit) mine is on the northeast side of the range, about 2 miles north of Bristol. (See fig. 10.) A settlement at this property, fair sized in 1912, was deserted except for a watchman in 1913, owing to the closing of the mine.

The Day shaft, about 2,000 feet south of the Jack Rabbit incline, is nearly vertical and reaches a depth of about 900 feet, with 12 levels. The largest ore bodies were stoped from the surface to the third level and between the fourth and the ninth levels. The lower stopes were on bodies of ore which occurred along a fissure bearing approximately N. 45° E. From the fourth to the sixth levels this fissure dips southeast, but below the sixth level it dips at an average of 50° NW. The maps of the mine indicate that little stoping was done below the ninth level.

The Jack Rabbit incline bears S. 50° W. and is approximately 1,200 feet long on the dip, which is 20°–30° S. In the upper levels, principally above the 200-foot level, there are a series of abandoned stopes on some replacement deposits in limestone along a fissure which strikes N. 10° W. and dips 60° W. The ore body was tabular along the strike of the fissure and ranged from a few inches to 4 feet thick. The rich ore, as shown by the exposures in a few pillars, was entirely a mixture of cerusite and horn silver that seems to have occurred next the hanging wall in a relatively thin seam.

The lower levels of this incline connect with the workings on the Day ore body, the 900-foot level of the Jack Rabbit incline connecting with the sixth level of the Day shaft. A long southwest drift, caved 1,750 feet from the incline, seems to follow the Day fissure, which is barren in much of the drift. The ore body is a replacement of dark crystalline limestones whose structure was not ascertained. The principal replacement seems to have taken place southeast of the more or less open fissure and to have extended into the limestones for 4 to 30 feet. It may be that the locus of this replacement occurred at the junction of the fissure trending N. 45° W. with the one trending N. 10° W., though this was not definitely determined.

The ore in the lower stopes visited consists for the most part of red and brown hydrous iron oxide that usually contains some copper and in many places carries appreciable amounts of copper carbonate. A large part of the red ore which does not appear to contain copper will give copper reactions if tested chemically. This low and medium grade ore contains in places bodies of soft black or dark-brown manganese oxide, which are said to constitute the higher-grade silver ore of the mine. They carry small stains of copper carbonate and here and there yellowish green stains that may be silver chloride. In the specimen of this manganiferous ore collected by the writer chemical tests revealed only a small amount of silver chloride.

OTHER PROPERTIES.

Malefactor claims.—The two Malefactor claims, a mile south of the Home Run mine, are along a belt of iron-stained breccia that strikes N. 15° W. and is from 20 to 50 feet wide. This breccia occurs about on the strike of the western fault. (See fig. 10.) The eastern fault zone is represented by a breccia about one-fourth mile east of the claims. No development work has been done on the claims, which were located in August, 1913, by Thomas Grover. Siliceous float that carries residual galena and is more or less stained with yellow earthy cerusite and occasionally by copper carbonates, indicates that some fair-sized quartz veins probably occur in the yellowish shaly limestone between the two fault zones.

Home Run mine.—The Home Run mine, near the south end of the Bristol district, is not far east of the large fault along the west side of the Highland Range. The main work is an incline 250 feet deep, which pitches 80° N. on a fissure that strikes N. 17° E. and dips 60° E. This fissure is tight in most places and the dense siliceous limestones which it cuts have not been replaced to any considerable extent. Some small bodies of copper carbonate and manganese oxide ore above the 200-foot level parallel the bedding of the limestones, which dips 5° N. At the bottom of the shaft crevices which enter the main fissure at an acute angle and dip 70° E. contain small lenses of copper carbonate ore. A quarter of a mile south of the shaft on the Home Run ground an eastward-trending vertical fissure, developed by an inaccessible shaft presumably about 50 feet deep and by a number of open cuts, carries siliceous copper ore.

Kismet properties.—The Kismet Mining Co. controls a group of seven patented claims and two locations lying east of the Day-Bristol properties. The principal development work is on the Adelaine claim (see fig. 10), where a shaft about 40 feet deep is sunk on a fissure that strikes N. 60° W. and dips 45°–60° N. Cerusite and copper carbonates replace the limestone along the fissure and bedding planes adjacent to it. In a tunnel 50 feet west of the shaft 12 to 18 inches of cerusite ore carrying a little copper lies parallel to the bedding, which dips 10° N. Most of this ore carries zinc carbonate and some of it is said to run as high as 19 per cent zinc.

On several of the other claims of this group there are shallow shafts and cuts on small replacement deposits carrying lead and copper carbonates.

Ida May claim.—The Ida May property, on the north slope of the ridge about a mile west-southwest of the Day mine, was not visited. The workings are said to show the replacement deposits similar to those common in the Bristol district.

NYE COUNTY.

LOCATION AND ACCESSIBILITY.

Nye County includes a considerable part of south-central Nevada. With the exception of the Las Vegas & Tonopah Railroad, which crosses its south end, there are no railroads within its boundaries. Tonopah, the county seat, is at the extreme western boundary, and is served by a branch of the Southern Pacific which leaves the main line at Hazen. Transportation facilities are not of the best in most of this region, where horse-drawn vehicles or automobiles are the only freight carriers. Fortunately, most of the roads in the northern part of the country, where the majority of the mining districts are located, are good desert roads. Travel is therefore not so difficult as it is in the more sandy desert region to the south. Tonopah to the west and Ely to the east are the main supply points of northern Nye County. Mail stages that start from these towns and from Austin, Lander County, serve most of the northern county. The southern part of Nye County has no stage service, as there are few settlements of any description in this region.

There are at least 33 mining districts in Nye County. Two of these, situated on the east side of Railroad Valley, in the Grant and Quinn Canyon ranges, in the northeast part of the county, were visited by the writer in October, 1913. (See Pl. I.) The Troy (Irwin Canyon) district (No. 19, fig. 1, p. 18) is about 30 miles and the Willow Creek district (No. 20) nearly 50 miles south of

Currant post office. Mormon Well, near Willow Creek district, is about midway between Ely and Tonopah, from either of which places it can be reached by stage.

TROY (IRWIN CANYON) DISTRICT.

LOCATION AND ACCESSIBILITY.

Troy and Irwin canyons (No. 19, fig. 1, p. 18), on the west side of the Grant Range, enter the east side of Railroad Valley about 30 miles south of Currant post office. A triweekly automobile stage runs between Ely and Currant. The road south from Currant is a good desert road except for a short sandy stretch about 15 miles south of Currant. The old Troy mine is near the head of Troy Canyon, 7 miles east of Railroad Valley. (See Pl. V.) Most of the properties in Irwin Canyon north of Troy are within 4 miles of the valley and are much easier of access than the Troy mine. The Vanderhoef claims are said to be in unsurveyed T. 6 N., R. 57 E., and are southeast of Bullwhacker and Willow springs, shown on the Land Office map of Nevada issued in 1908.

TOPOGRAPHY.

Troy and Irwin canyons are two westward-draining channels that are about $2\frac{1}{2}$ miles apart and are separated by a rugged spur of the Grant Range. Troy Canyon rises approximately 3,000 feet in the 7 miles between its mouth and the old Troy mine, whose elevation is 8,650 feet by barometer. The summit of Grant Range, somewhat over a mile east of the Troy mine, is nearly a thousand feet higher than the mine. Irwin Canyon rises about 1,500 feet from its mouth to Vanderhoef camp, a distance of 4 miles. A characteristic of both canyons is the steep rise in the lower 2 miles of their courses.

The stock of quartz monzonite at the edge of Railroad Valley (Pl. V) rises as a prominent knob above the valley and is joined to the main range by a pass at least 600 feet lower than the summit of the hill at the western end of the ridge. The divide north of the lower part of Irwin Canyon is relatively low, and north of it a broad arm of Railroad Valley extends eastward for several miles toward the axis of the Grant Range. South of the ridge south of Troy Canyon a broad deep notch extends southwest toward Garden Valley, which separates the Grant and Quinn Canyon ranges.

Railroad Valley, lying west of the Grant and White Pine ranges, is about 80 miles long and 16 miles wide. It is practically level and contains a large playa basin whose center is about 20 miles southwest of Currant post office. Gently sloping outwash plains and cones extend from the mountains for several miles toward the axis of the valley. Along the east edge of the valley between Troy Canyon and Currant a number of large springs, whose waters are used to irrigate ranches of considerable size, issue from the outwash gravels not far from the break of the mountains.

GEOLOGY.

In Troy and Irwin canyons a considerable thickness of shaly limestones overlain by white quartzite, which is in turn overlain by massive bedded limestones, have been intruded by a large stock of quartz monzonite, which forms a prominent mountain at the edge of Railroad Valley. (See Pl. V.) The extent of the igneous rock is not known, though it is believed that it does not continue many miles northeast of the area shown.

SEDIMENTARY ROCKS.

The oldest sedimentary rocks are black and drab thin-bedded lime shales and argillites. They are best exposed in Troy Canyon between the old town and the Troy mine. The rocks have been folded into a number of shallow anticlines and synclines and in the vicinity of the Troy mine into a rather tight anticline. (See Pl. V.) In this fold the nearly black calcareous argillites have been altered to black micaceous schists. The thickness of this formation was not worked out but is roughly estimated to be about 3,000 feet.

The quartzite that overlies the shale east of the Troy mine and outcrops in small exposures at the mouth of Troy Canyon and on the divide north of Irwin Canyon is a fine-grained vitreous white rock about 200 feet thick, which seems to be very persistent.

. The limestones overlying the quartzites were not seen in place except at the mouth of Troy Canyon, where they are evidently altered by the intrusion to crystalline marble-like rocks of light color. The cliffs east of the Troy mine are composed of thick beds of dark blue-gray semicrystalline dolomitic limestone. The limestones north of Irwin Canyon are in comparatively thin beds and where seen near the intrusive contact they are crystalline.

The white quartzite of the series exposed in this region is believed to be the equivalent of the Ordovician Eureka quartzite. If this is the case, the underlying shaly limestones should be referred to the Ordovician Pogonip limestone, and the overlying massive beds to the Lone Mountain limestone, also Ordovician, of the section at Eureka, Nev., described by Hague.[1]

Spurr[2] evidently had little hesitancy in correlating these formations with the Ordovician section of Eureka.

[1] Hague, Arnold, Geology of the Eureka district, Nev.: U. S. Geol. Survey Mon. 20, pp. 47–62, 1892.
[2] Spurr, J. E., Descriptive geology of Nevada south of the fortieth parallel and adjacent portions of California : U. S. Geol. Survey Bull. 208, pp. 67–71, 1903.

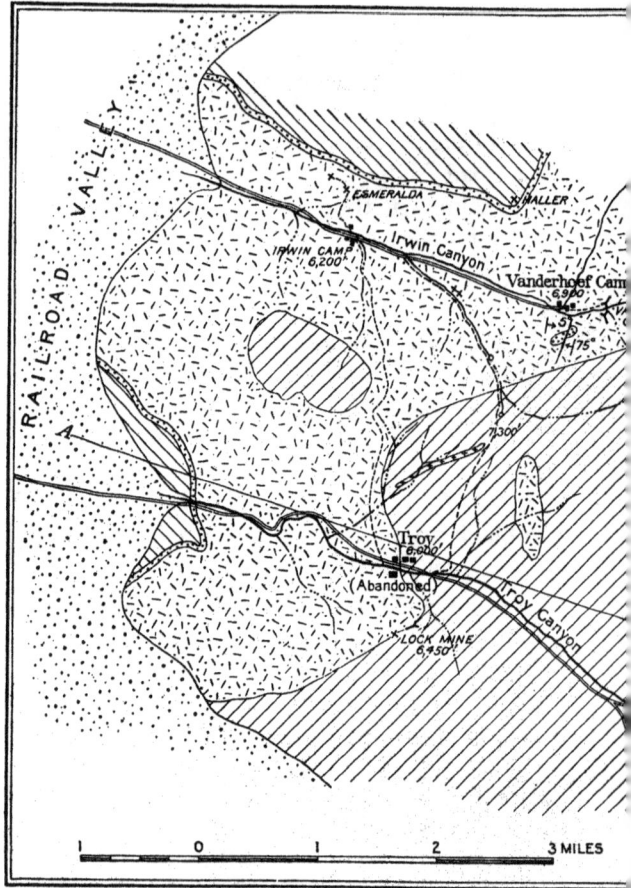

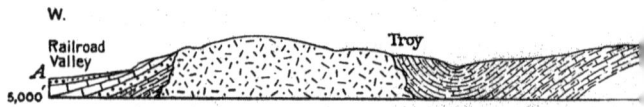

SKETCH MAP OF THE TROY (IRWIN CANYO

LEGEND

SEDIMENTARY ROCKS

Gravel and sand — QUATERNARY

Limestone, blue, massive bedded

White quartzite

Thin-bedded shaly limestone, schistose in places; some argillite — ORDOVICIAN (?)

IGNEOUS ROCKS

Quartz monzonite

Rhyolite — PROBABLY CRETACEOUS OR EARLY TERTIARY

Fault

Strike and dip $\times 75°$

Mines and prospects

Trail

N

G R A N T   R A N G E

TROY MINE
8,650'

A'

E. A'

TROY MINE

5,000'

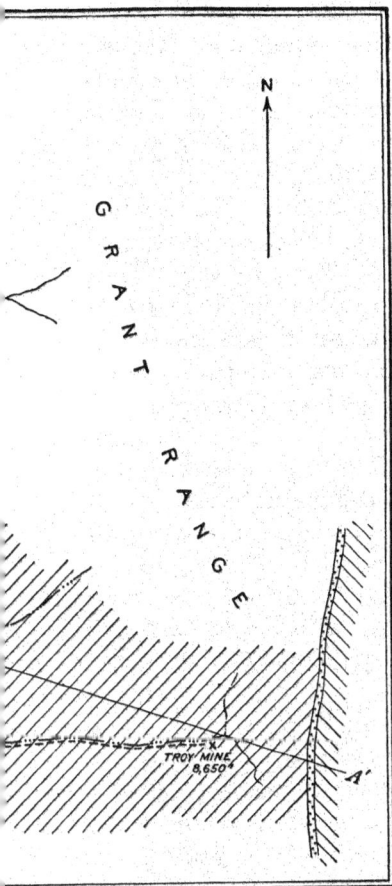

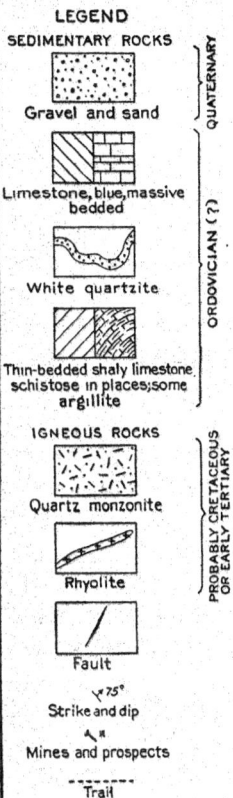

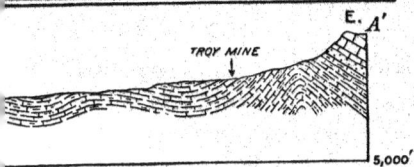

TRICT, RYE COUNTY, NEV.

Quartz monzonite.—A stock of quartz monzonite forms the mountain northwest of Troy and extends northeast for an unknown distance beyond the area shown on Plate V. The typical rock is light gray, rather coarsely crystalline, and of very uneven grain. Most of the mass is inequigranular rather than porphyritic, though decidedly porphyritic facies were seen. Movement along the southeastern edge of the stock has produced gneissic banding. Biotite and muscovite, together with quartz and two kinds of feldspar, are visible in hand specimens of the typical quartz monzonite. The quartz and feldspar are more abundant than the micas. In thin section andesine, quartz, orthoclase, muscovite, biotite, apatite, and magnetite, named in the order of their abundance, are seen to form the rock. In some sections the micas are more abundant than orthoclase. In the gneissic portion of the stock the banding is due rather to crushing than to rearrangement of the constituents. In one specimen, however, the micas tended to have their plates arranged parallel to the banding. In not a few places, particularly in the south fork of Irwin Canyon, near the contact, there are large masses of the quartz monzonite whose constituents are of pegmatitic size.

Near Irwin camp a body of dark rock seems to be intrusive into the quartz monzonite, but there is little question that it is a differentiation product from the same magma. This is a fine-grained gray equigranular rock consisting essentially of orthoclase, andesine, quartz, greenish-brown hornblende, colorless pyroxene and abundant apatite and magnetite.

The western and southern contacts of the stock appear to stand at steep angles and at only a few places was there any indication of the development of lime silicate minerals along the contact, but the limestones and lime shales have been altered to crystalline rocks. At the Lock mine, south of Troy, which is directly on the contact, no lime silicates were noted. The northern contact, north of Irwin Canyon, dips 15° N.

Rhyolite.—On the north side of Troy Canyon, about a mile north of Troy, a narrow dike of a purplish-red glassy rhyolite trends northeast about parallel to the edge of the stock. Phenocrysts of orthoclase, quartz, and biotite composing about one-fourth of the rock, are set in a deeply iron-stained glass in which there are microscopic spherulites that seem to be quartz. Float of somewhat similar rock was seen at the mouth of the south fork of Troy Canyon.

ORE DEPOSITS.

HISTORY.

The Troy ore deposit, covered by the Gray Eagle and Troy claims, was discovered in 1868, according to Raymond,[1] and was bought by a Yorkshire, England, company in 1869. This company built a 20-stamp mill with Stetefeldt furnaces at Troy, 5½ miles west of the mine. Raymond reported that the lode conformed to the stratification of dark argillaceous shale and consisted of zinc blende with some chalcopyrite and pyrite in a white quartz gangue. The English company is reported[2] to have spent about $500,000 and to have shut down permanently in 1872, after a two years' trial, as the silver content of the ore was too low to pay the mining and treatment costs.

The Lock mine was worked by the same company, and has been prospected by various persons since the English company shut down.

The Esmeralda claims, near Irwin camp, were located by F. L. Irwin in 1905. Some work has been done each year to develop the ground, but so far as could be learned, no ore has been shipped. E. E. Vanderhoef, of Ely, Nev., located the 13 claims of the Vanderhoef group in August, 1912, since which time development has been under way.

THE PROPERTIES.

Troy mine.—The old abandoned workings on the Troy and Gray Eagle lodes are at the head of Troy Canyon, about 5½ miles east of the camp and mill, at a barometric elevation of 8,650 feet. (See Pl. V.) The main work was through an incline shaft, said to be 500 feet deep, that was under water in October, 1913. A tunnel about 100 feet below the collar of the shaft was caved near the mouth but is shown by the dump to be of considerable length. Other tunnels run north and south from the gulch on the lode. The northern tunnel could be entered for a short distance. These workings are on large lenslike bodies of white quartz which dip 40°–50° E., parallel to the thin-bedded black lime shales on the east limb of an anticline shown in cross section on Plate V. Some particularly black beds are argillaceous lime shales that have been made schistose and that strike N. 50° E. and dip 45° SE. The white quartz lenses are irregular in detail and, where seen, vary from a few inches to 20 feet thick. The hanging wall seems to be much more regular than the footwall in the lens opened by the tunnel on the north side of the canyon.

---

[1] Raymond, R. W., Statistics of mines and mining in the States and Territories west of the Rocky Mountains, for 1873, p. 227, 1874.

[2] Whitehall, H. R., Nevada State Mineralogist Fifth Biennial Rept., for 1873–74, pp. 76–77, 1875.

Much of the quartz on the dump is barren, though some specimens showing the light-blue and green stains of copper were found. Similar ore seen about the old mill at Troy is said to carry silver, but chemical tests failed to show the presence of that metal in appreciable amounts. Some ore obtained in a 60-foot winze from the north tunnel is white quartz with small amounts of black sphalerite and a little copper-bearing pyrite. The two sulphides are intergrown. Tests of the pure zinc sulphide show that it contains traces of copper and considerable iron as well as zinc. It is reported that galena was found with zinc sulphide in the main shaft, though none was seen on the dump.

Lock mine.—The workings of the Lock mine are on the east and west sides of a low saddle on the ridge about three-fourths mile south of Troy. (See Pl. V.) East of the saddle a tunnel has been driven west-southwest for about 500 feet on the contact between the quartz monzonite and lime shales. The lime shales have suffered no contact metamorphism in this vicinity. The contact is nearly vertical and considerable postintrusion movement has occurred along it. The quartz monzonite near the contact is somewhat gneissic and is cut by a great number of small faults along which the movement, as evidenced by striæ, was nearly horizontal. These small faults strike N. 40° E. or N. 60° W. The tunnel follows the fault planes and in consequence is very tortuous. At one place it follows a 4-foot seam of slightly iron-stained bull quartz that cuts the quartz monzonite, and is said to carry some gold. West of the saddle several short inclines and tunnels have been driven in a lenslike body of iron-stained quartz that lies parallel to the bedding of the lime shales about 100 feet south of the intrusive contact. The shales strike N. 20° E. and dip 25° E. The lens seems to pinch out in a distance of 100 feet on the dip and is not continuous along its strike for more than 300 feet.

Vanderhoef claims.—The Vanderhoef claims are on both sides of Irwin Canyon about 4 miles from the edge of Railroad Valley. There are 13 claims in the group, but development is largely confined to the Lilly and Garden claims north and south of the canyon near the camp. (See Pl. V.) Two tunnels, 180 feet and 60 feet long, are on the south side of the canyon, and other tunnels, 100 feet and 50 feet long, are on the same lode north of the stream. The lode which averages 20 feet wide, consists of white quartz that is somewhat iron stained at the surface and contains some galena, sphalerite, and pyrite at a shallow depth. The sulphides are not abundant and are irregularly spotted throughout the gangue. Pockets of cerusite found at the surface are said to have assayed 60 ounces a ton in silver. The lode occurs in a small lens of micaceous shaly limestone, engulfed in the quartz monzonite, which strikes about N. 70° W.

and dips south at medium angles. On account of the steep slope of the canyon sides, the croppings on either side of the creek were thought to be different bodies of quartz, but this does not seem to be the case.

Haller property.—The Haller claims are near the top of the hill northwest of Vanderhoef camp in Irwin Canyon. In this locality the contact of the shale, quartzite, and quartz monzonite dips north and west at low angles. A very thin bed of white quartzite lies between the intrusive rock and the schistose lime shales in which the ore occurs. The ore body, a thin tabular replacement of the shales parallel to the bedding, is developed by a short tunnel and flat stopes. At the back of the stopes a fault, along which the movement was nearly horizontal, has cut off the ore. This fault strikes N. 85° E. and dips 85° N. The ore minerals are galena, sphalerite, and some pyrite. There has been very little oxidation of the ore, though some thin crusts of anglesite and cerusite were seen lining cracks in the sulphide ore.

Irwin group.—The five claims of the Esmeralda group, located in 1905 by F. L. Irwin, are on the north side of Irwin Canyon, about 2 miles east of Railroad Valley. (See Pl. V.) These claims cover a series of veins cutting quartz monzonite.

The No. 1 vein, about one-fourth mile north of Irwin camp, strikes N. 75° E. and dips 75°–80° N. in places but is nearly vertical in others. It is developed by a shaft with an east drift 150 feet at the 80-foot level, below which water stands in the shallow sump. Above this level the vein is seen to vary from a few inches to a maximum width of 3 feet. It consists of white quartz, much pitted and stained with iron near the surface, that is said to contain free gold. At the 80-foot level large bunches of somewhat oxidized pyrite are irregularly distributed throughout the quartz, and occasionally small masses of a mixture of dark sphalerite and galena are seen. The vein is separated from both walls by one-half to 2 inches of gouge that is said to carry gold. The quartz monzonite is rusty and sericitized for 2 feet on either side of the vein.

The No. 2 vein, opened by a 40-foot shaft about 300 feet northwest of the No. 1 shaft, consists of 2 feet of quartz and 3 feet of crushed, altered, iron-stained quartz monzonite. Below the altered country rock there is a clay gouge. Pyrite and minor amounts of sphalerite and galena are sparingly distributed throughout the quartz and altered rock near the bottom of the incline.

A tunnel a short distance northeast of the No. 2 shaft starts in altered quartz monzonite and runs north into sedimentary rocks. About 3 feet of quartzite lies above the fault that marks the contact, which dips 15° N. Above the quartzite lies crystalline limestone, which is also exposed in the low hills for a considerable distance to the north.

Near Irwin camp a small, poorly exposed body of altered diorite clearly intrudes the quartz monzonite.

Other prospects.—South of Irwin Canyon several prospects on quartz veins in the quartz monzonite are very similar to those developed at the Esmeralda. Many of these veins at the surface are not much iron stained and few are developed sufficiently to show sulphides. As a rule the veins in the quartz monzonite strike N. 50°–75° E., about parallel to the most prominent jointing of the stock. Some of the veins are frozen to the wall rock in some places, but most of them are separated from it by gouge. The quartz monzonite is bleached and sericitized for a few inches on both sides of the veins, the zone of alteration being about as wide as the quartz filling of the vein.

It is reported that some prospects in the massive limestone above the quartzite at the head of Troy Canyon are on bodies of lead ore said to carry silver and gold. Some ore, said to have come from these prospects, consists largely of cerusite but shows also kernels of galena and a little anglesite.

WILLOW CREEK DISTRICT.

LOCATION AND ACCESSIBILITY.

The Willow Creek district of northeastern Nye County (No. 20, fig. 1, p. 18) is in the foothills on the northwest side of the Quinn Canyon Mountains, near the south end of Railroad Valley. According to W. B. Colwell, a deputy mineral surveyor of Ely, Nev., the springs which constitute the main water supply of the district are about 2.7 miles S. 12° 30′ W. of the southeast corner of T. 5 N., R. 55 E. (see fig. 13), in Willow Canyon, about 2 miles from Railroad Valley. The center of the district is approximately 12 miles north of latitude 38° and 15 miles east of longitude 116°.

Willow Creek is about 90 miles southwest of Ely, on the Nevada Northern Railway, and about the same distance from Tonopah in the western part of the State. From either of these towns it may be reached over a fair desert road. In 1913 an automobile stage made daily trips between Ely and Currant post office, on Kern Creek at the north end of Railroad Valley. A somewhat irregular stage service between Currant and Tonopah follows the east side of Railroad Valley south to Mormon Well, where it turns westward and goes to Tonopah by way of Twin Springs.

TOPOGRAPHY.

The Willow Creek prospects are in the lower foothills of the range in westward and northwestward draining canyons. Willow Canyon is in lime shales except in its lower part, where it cuts a small stock

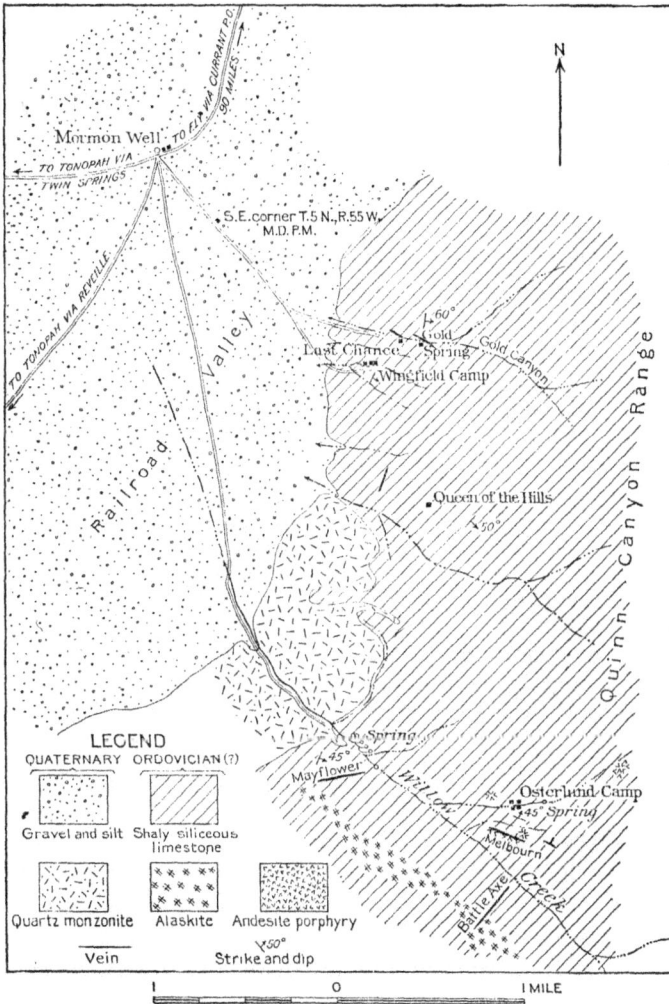

FIGURE 13.—Sketch map of the Willow Creek district, Nye County, Nev.

of quartz monzonite. Its middle portion leads through low rounded hills, but in the monzonite area it is deep and steep sided. Its head is in dense massive limestones and is rugged. Gold Canyon is in thin-bedded siliceous mica schists and has steep cliff-like walls.

42680°—Bull. 648—16——10

The prospects are at elevations of 6,600 to 7,000 feet, by barometer, and are not over 1,500 feet above the general level of Railroad Valley. East of the mines the main range rises to 9,000 feet above sea level.

GEOLOGY.

SEDIMENTARY ROCKS.

The veins are found in thin-bedded yellowish and dark-gray shaly limestones which grade into massive dark-gray limestones. They strike approximately north-northeast and dip rather steeply east. The lower shaly beds north of Gold Canyon are schistose and contain abundant light-brown biotite. Above the shales are thick-bedded dark-gray limestones, which are overlain by about 200 feet of vitreous white quartzite thought to be the equivalent of the Ordovician Eureka quartzite. The lower schistose shales are thought to be of Upper Cambrian age,[1] possibly the equivalent of the Dunderberg shale of the Eureka section, and the overlying lime shales and limestones are considered to be the equivalent of the Ordovician Pogonip limestone.

Spurr,[2] in discussing this region, says:

In the foothills at the north end of Quinn Canyon Range * * * an exposure of rusty-brown shaly limestone was found * * * from which fossils were collected. They were determined by Mr. Walcott as Cambrian. * * * On the steep west face of the northern end of the Quinn Canyon Range the mountains near the base consist of massive, often shaly, dark-blue to gray-blue limestone, much broken and veined as a consequence of granite intrusions. On account of the alteration the organic remains obtained from this limestone are not identifiable. Six hundred or 800 feet above the base of the limestone, as exposed, comes about 200 feet of hard white vitreous quartzite, which one at once recognizes as probably the Eureka formation.

IGNEOUS ROCKS.

Quartz monzonite.—At the mouth of Willow Creek a low rounded hill of quartz monzonite intrudes the sedimentary rocks. The north, west, and south contacts of this stock are nearly vertical, as is much of the eastern contact, but a long tongue of the limestone extends over the stock from the east side. (See fig. 13.) The rock is light gray and inequigranular and weathers in tawny or buff colors. Here and there it contains rather large white quartz and a few orthoclase phenocrysts, and in many places carries biotite and hornblende of megascopic size. In thin section the rock is seen to carry nearly equal amounts of zonally built andesine and orthoclase. The feldspars are more abundant than quartz, brown biotite, and light-

[1] Hague, Arnold, Geology of the Eureka district, Nev.: U. S. Geol. Survey Mon. 20, pp. 41, 47–57, 1892.

[2] Spurr, J. E., Descriptive geology of Nevada south of the fortieth parallel and adjacent portions of California: U. S. Geol. Survey Bull. 208, p. 69, 1903.

green hornblende. Apatite and magnetite are common, though not abundant, accessory minerals, and a minor amount of rutile is present. The plagioclase feldspars are only slightly sericitized, but some of the other minerals show alteration.

Along the south contact in Willow Canyon the shaly limestones have been altered to lime silicate rocks. The contact-metamorphosed zones are usually narrow, in no place exceeding 15 feet in width. A few blocks of limestone engulfed in the quartz monzonite have also been metamorphosed, though some only slightly so. The material from the contact zone in Willow Canyon is roughly banded, carrying similar minerals in the different bands but in different proportions. Iron oxide, bluish green amphibole, and brownish-yellow garnet are the most abundant minerals, but quartz and calcite are present in places in considerable amounts.

Andesite porphyry.—Near Osterlund camp in Willow Creek (see fig. 13) several small areas of a dark-gray to black igneous rock intrude the sediments. This rock, called diorite by the prospectors, proves to be a glassy porphyritic andesite. Large well-developed crystals of andesine-labradorite and a yellow amphibole are the phenocrystic minerals. They are set in a dark glass crowded with specks of magnetite and a few plagioclase microlites. The amphibole is largely altered to calcite. A small amount of calcite is developed parallel to the cleavage cracks of some of the plagioclase crystals.

Alaskite porphyry.—On the south side of Willow Creek about 200 feet above the canyon bottom an irregular dike of fine-grained white porphyry shows slightly iron-stained croppings. The dike varies from 20 to 100 feet in width, strikes about parallel to the canyon, and seems to stand about vertical. The rock consists essentially of microscopic grains of orthoclase and quartz, in some places carrying phenocrysts of quartz. In the specimens collected the orthoclase is altered to sericite.

Rhyolite.—On the summit of the south side of Willow Creek the edge of a flow of dark-gray porphyritic rhyolite is seen. Similar rock is said to cover a considerable area of the mountains south of Willow Creek, being particularly well developed in Big Creek Canyon.[1] A thin section of this rock shows phenocrysts of quartz, orthoclase, biotite, and oligoclase set in a dark glass base.

ORE DEPOSITS.

HISTORY.

The oldest claim in this camp, the Rustler, was located in June, 1911, by Charles Sampson and David Jenkins. These men located

---

[1] Spurr, J. E., Descriptive geology of Nevada south of the fortieth parallel and adjacent portions of California : U. S. Geol. Survey Bull. 208, pp. 72–73, 1903.

five adjoining claims during 1911 and 1912, and the group was purchased by George Wingfield in April, 1913. It is reported that about 40 tons of ore shipped from this property carried about $8 in gold and $65 in silver a ton. The excitement in the district in 1913 was caused by the discovery in April, 1913, by Steve Papas and W. E. Blackwell, of free gold in the Melbourn vein. At the time of visit the four Melbourn claims were bonded to Mr. Osterlund and associates, of Ely, Nev. A 500-pound shipment of high-grade free-gold ore made in October, 1913, is said to have returned $39.40 a pound.

OCCURRENCE AND CHARACTER OF THE ORES.

At least two types of ore deposits occur in the Willow Creek district, but the development in the fall of 1913 was hardly sufficient to permit a complete understanding of the deposits.

One type is represented by the Melbourn, Battle Axe, and Mayflower in Willow Creek and by the Last Chance and Gold Spring in Gold Canyon. These are quartz veins carrying free gold closely associated with a little talc and calcite. The white quartz, without talc, is said to contain little or no gold. So far as development shows, the original minerals of these veins were pyrite and chalcopyrite. The veins cut across the bedding of the sediments and were deposited in open fissures. Movement since the deposition of most of the veins has crushed the quartz. The talc and free gold are found in the crushed belts of quartz rather than in the parts of the veins which were not crushed. As examined in thin sections the uncrushed quartz is seen to be strained and to be crowded with minute inclusions which seem to be solid. The croppings of the veins are nowhere highly iron stained and do not seem to have been strongly mineralized.

The second type of deposit is best exposed at the Rustler mine but is also seen at the Queen of the West. These deposits appear to be replacements of limestone along bedding planes near fissures. Arsenopyrite and argentiferous tetrahedrite appear to have been the principal minerals originally deposited. The ores so far developed are oxidized to masses of a copper pitch ore, limonite, and copper carbonate and silicate. Free silver is seen in much of the ore from the Rustler, and it is said that silver chloride has been found. The ore is valuable chiefly for silver, but carries small amounts of gold and copper.

THE PROPERTIES.

Gold Spring group.—The Gold Spring group of five claims is in Gold Canyon (the northernmost canyon shown on fig. 13). The principal development work is in the canyon bottom at a barometric elevation of 6,850 feet. In this vicinity the dark lime shales are rather siliceous, have in many places been changed to mica schists, and are cut in all directions by veinlets of calcite and quartz, which have a prevailing north-northeast strike and steep east dip. A vertical vein that is followed by a short tunnel running N. 60° W. on the north side of the canyon appears to be the continuation of a vein striking N. 45° W. and dipping 52° NE., that is exposed in shallow workings on the south side of the canyon. In the south workings a vein striking N. 80° E. and dipping 50° N. intersects the west-northwest vein. Both veins are composed of white quartz and in places are frozen to the walls, from which many narrow quartz veinlets enter the main fissure. Some of the quartz is fine granular with drusy vugs, though most of it is in fairly large crystals developed perpendicular to the vein walls. A few casts of pyrite and iron oxide pseudomorphs of pyrite were seen, and in one specimen the pyrite was not completely oxidized. The quartz has been crushed by later movement, which is indicated by a thin selvage on the footwall of the vein as well as by the crushed quartz. Along the joints produced by this crushing free gold and green talc have been deposited. The richest ore invariably contains talc.

Last Chance tunnel.—The Last Chance tunnel, on claims belonging to William Wittenburg and Dr. Weller, of Goldfield, is 600 feet west of the Gold Springs work. It is a crosscut running N. 10° W. for 175 feet, following the bedding of micaceous slates, which dips 75° E. This tunnel had not yet cut the main vein, which is exposed in some pits and open cuts on the hill. This vein strikes N. 70° W., is nearly vertical, and consists of white quartz with some green talc. The tunnel cuts a 2 to 6 inch vein of white quartz that strikes N. 65° W. and dips 85° S. It is frozen to the hanging wall but is separated from the footwall by a paper-thin gouge.

Willow Creek Mining Co.'s claims.—The Willow Creek Mining Co., controlled by George Wingfield, is operating a group of six claims in Saversburg Canyon about 2½ miles SE. of Mormon Well. (See fig. 13.) The main work is on the Rustler claim, which was located in June, 1911. The development consists of a shaft about 50 feet deep and a 120-foot tunnel which cuts the shaft 40 feet from the mouth and 20 feet below the collar. A crosscut tunnel has been started to intersect the ore body at a depth of 125 feet. The country rock is a dark-blue limestone interbedded with shaly limestones. The formations strike N. 20° E. and dip east at low angles. Two fissures intersect at the shaft, one striking north and dipping 45° E. and the other striking N. 25° W. and dipping 85° E. Along the latter there is about 2 feet of crushed ore stained with iron and copper. The tunnel exposes a flat westward-dipping fissure, along which some ore has been deposited for about 115 feet. At the end of the tunnel a northward-trending fissure which dips 75° E. is followed by 8 inches to 2 feet of ore but is pinched and barren at the end of a short drift. The main ore body is in the nearly vertical fissure cut near the mouth of the tunnel. What is apparently a lens of ore is about 6 feet thick and 15 feet long at the tunnel level and about 18 inches thick at the bottom of the shaft. The material is all oxidized and carries considerable limonite, with some copper carbonates and silicates. Some of the ore from the bottom of the shaft consists of a partly oxidized mixture of arsenopyrite and a dark, nearly black submetallic mineral which contains silver, copper, iron, antimony, and a little sulphur. A polished surface of the sulphides shows the dark material to be a mixture of a soft light-gray metallic mineral and a nearly black nonmetallic substance. Both are in too fine distribution to be accurately determined, but it is thought that the gray mineral is a silver-bearing tetrahedrite and the nonmetallic substance an oxide that contains copper and possibly silver. This ore has been considerably crushed and carries limonite and native silver deposited in its joints. It is believed that the original deposit was a fissure replacement of the limestone rather than a true vein.

Queen of the West prospect.—The Queen of the West deposit is developed by two pits. It seems to be a replacement of limestone immediately above shaly limestone. The ore is an iron-stained siliceous material that carries a little copper carbonates and silicate and is said to contain silver.

About a quarter of a mile northwest of the Queen of the West prospect (see fig. 13) a 2-foot vein strikes N. 20° E. and dips 45° E., about parallel to the bedding of the shaly limestones. The footwall of this vein is a narrow dike of porphyritic quartz monzonite. The vein, frozen to the hanging wall, consists largely of arsenopyrite and quartz, but evidently carries some tetrahedrite, as the ore is copper stained.

Mayflower claim.—The Mayflower vein on the south side of Willow Creek near the springs (see fig. 13) strikes N. 80° E. and dips 85° S. It is developed by some shallow pits and cuts for about 300 feet. It cuts directly across the bedding of greenish-gray lime shales which strike N. 20° W. and dip 45° E. The vein varies from 2 to 3 feet in width and is separated from both walls by gouge. Next to the hanging-wall slip there is a broad band of white quartz with well-developed crystals, which stand perpendicular to the wall. In the quartz there are a few small bodies of copper-bearing pyrite, largely altered to a black resinous mineral resembling copper pitch ore. On the footwall there is a band of coarsely crystalline white calcite that is apparently unmineralized.

Melbourn mine.—The Melbourn vein, about a mile above the spring on the north side of Willow Canyon, was developed by a drift tunnel 150 feet long, with a 50-foot winze about 30 feet from the tunnel mouth. The vein strikes N. 67° W. and dips 40°–70° S. It varies from 10 to 18 inches in width and consists of white quartz that has been crushed and carries green talc along the joints. Some fragments of altered lime shale country rock are included in the quartz. Near the mouth of the tunnel a dike of glassy andesite porphyry forms the hanging wall of the vein but is apparently cut by it. The vein is separated from the dike by 1 to 2 inches of gouge but is frozen to the shale walls, though at a number of places slip planes about parallel to the vein lie in the walls outside of the quartz. Calcite accompanies the quartz and talc in the 60-foot winze. Specks of limonite in this ore have cubical forms as if the oxide was derived directly from pyrite. Native gold is the valuable constituent of the ore. Some beautiful specimens of gold in the talc were seen at Ely and the mine. Most of the gold is associated with the green talc, though some is said to occur in the slightly iron-stained quartz. A northward-trending fault, which dips 45° E., cuts the vein 70 feet from the mouth of the tunnel. East of this fault the vein has been displaced 4 feet to the north, indicating a reverse movement along the fault.

Battle Axe claims.—The three Battle Axe claims are on the south side of Willow Creek, south of the Melbourn ground. (See fig. 13.) The vein exposed in the east claim strikes N. 42° E. and dips 60° SE. and is lodelike, consisting of a main vein 10 inches wide and several narrow subparallel sugar-quartz stringers. The vein has been traced for about 300 feet along its strike. It cuts the lime shales and also a large body of alaskite porphyry which is intrusive into the shales. (See fig. 13.) In a number of pits greenish talc occurs with the quartz, but so far as seen it is not very abundant. The quartz is said to average about $6.40 a ton in gold and silver.

WHITE PINE COUNTY.

LOCATION AND ACCESSIBILITY.

White Pine is the middle county of the east tier in Nevada. (See fig. 1, p. 18.) Ely, the county seat, is near the center of the county in Steptoe Valley, which for many years was one of the main routes of travel from the Southern Pacific to the southeastern part of the State and is now followed by the Nevada Northern Railway, which leaves the main line of the Southern Pacific at Cobre. Many of the mines of White Pine County were opened in the sixties, and the main routes of travel have been established for many years. The Overland Stage Route passed through the northern part of the county by way of Tippett, Schellbourne, and Cherry Creek. (See Pl. I.) The roads in this county are almost all good, and the numerous stage lines that radiate from Cherry Creek and Ely make all parts of the county easily accessible. Most of the mountains are of sufficient height to be well watered, and as a consequence the great majority of the valleys are under cultivation in many places.

Nine of the 12 mining districts in this county were visited in 1913, and are described in the following report by ranges. (See Pl. I and fig. 1, p. 18.) The most western district visited was the Bald Mountain, at the south end of the Ruby Mountains, in the northwest corner of the county. In the Egan Range the districts from north to south are Cherry Creek (Egan Canyon), Hunter, Granite (Steptoe), and Ward. In the Schell Creek Range, east of Steptoe Valley, the districts are Aurum (including Schellbourne, Siegel, Old Aurum, and Muncy Creek), Duck Creek, and Taylor. The most eastern district visited was the Kern, at the south end of Kern Mountains in the northeastern part of the county.

BALD MOUNTAIN DISTRICT.

LOCATION AND ACCESSIBILITY.

Bald Mountain district (No. 22, fig. 1, p. 18) is in a pass between North and South Bald mountains, the former peak being 7 miles south of Ruby Pass (formerly called Hastings Pass) and 12 miles south-southwest of old Fort Ruby, a station on the Overland Stage route at the south end of Ruby Lake. (See Pl. I.) The center of the district is about 6 miles south of latitude 40° and 4 miles west of longitude 115° 30'. (See fig. 14.) It is at the south end of the Humboldt, or, as it is now called, the Ruby Range.

Joy post office, in Water Canyon, is served biweekly from Eureka, 56 miles southwest of the camp. Freight is usually brought into this country from Elko or Halleck on the Southern Pacific, about 80 miles north of Joy, though some supplies are obtained from Currie, Cherry Creek, and Ely on the Nevada Northern Railway, 40 to 50 miles to the east.

ECONOMIC CONDITIONS.

Springs are not numerous in the district, though water rises in Water Canyon above Joy, below the Copper Basin mine on the east side of the divide and in a number of small seeps at various places in the gulches north of Joy. Most of the shallow shafts, particularly in the area of intrusive rock near Joy, are under water, and it would seem that a fair water supply could be obtained by pumping from any of them. Wood is fairly abundant and some of the piñon trees in the vicinity are of sufficient size for mining timbers. It is said that good timber can be obtained a few miles southwest of the mines.

There are no mills in the district, and the long freight haul has greatly hindered development. In September, 1913, very little work was being done. The mine workings at the Copper Basin, probably the most extensive at any one place, could not be entered.

TOPOGRAPHY.

The low hills in which the Bald Mountain district is situated form the southern end of the Ruby Range. A low, even-crested divide

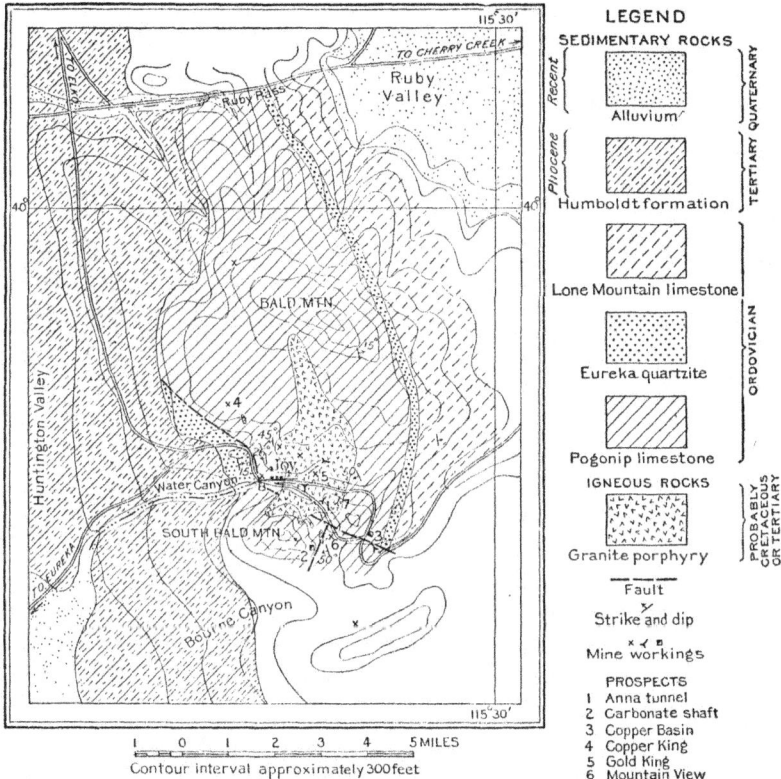

FIGURE 14.—Sketch map of the Bald Mountain district, White Pine County, Nev.

continues southward for several miles to the White Pine Range. From Huntington Valley on the west side the mountains rise in a gradual even slope to Bald Mountain. The northeast flank of that peak appears to be more abrupt, but eastward a moderately high plateau country extends for some distance, joining the south end of the Ruby Hills southeast and east of Ruby Lake. (See Pl. I.)

Barometric readings give the elevation of Joy as 7,400, South Bald Mountain as 9,000, Bald Mountains as about 9,400 feet, and Copper Basin Pass as approximately 7,800 feet above sea level.

GEOLOGY.

DOMINANT ROCKS.

As pointed out by Hague,[1] Ruby Range south of Harrison (Ruby) Pass consists " of a single series of limestones conformably underlain by quartzites, the latter appearing along the western base, while the entire summit and eastern face presents only heavy massive limestone," which Hague considered of Devonian or Carboniferous age. On the atlas accompanying his report, west half of Sheet IV, Bald Mountain is shown as intrusive granite. During the brief visit on which this report is based only the vicinity of the mines was studied, yet it seems that the " granite " occupies a comparatively small area south of Bald Mountain. (See fig. 14.)

SEDIMENTARY ROCKS.

From the few data obtained relative to the stratigraphic and structural geology, it would appear that Bald Mountain is a low anticline, whose western limb has been eroded so that the older beds are exposed on the western flank. The thin bed of vitreous white quartzite mapped by the geologists of the Fortieth Parallel Survey along the eastern side of the mountain is believed to correspond to the Ordovician Eureka quartzite of the Eureka district, which is 50 miles southwest of Joy in the Diamond Range across Huntington Valley. The underlying light-colored limestones forming the main body of Bald Mountain are considered to be the equivalent of the upper part of the Ordovician Pogonip limestone.[2] The lower part of the limestones overlying the supposed Eureka quartzite is believed to be the equivalent of the Lone Mountain limestone.

The main body of the " granite " stock has been intruded in this limestone along a fault striking N. 60° W., which is best seen at the Copper Basin workings. This fault separates the light-colored limestone on the north from a body of dark-gray to black fine-grained rough-textured dolomitic limestones which form most of South Bald Mountain, where they are interbedded with some dark quartzites and brownish shales. The latter are seen in the south fork of Water Canyon about three-quarters of a mile south of Joy. This series is considered to be equivalent to the lower part of the Pogonip limestone.

Huntington Valley is deeply filled with partly consolidated, horizontally bedded sands, which toward the mountains grade into fine sandy conglomerates that dip gently toward the valley. Deep washes are cut into these beds, which are described by King[3] as the Hum-

[1] Hague, Arnold, U. S. Geol. Expl. 40th Par. Rept., vol. 2, pp. 528–532, 1877.

[2] Hague, Arnold, Geology of the Eureka district, Nev.: U. S. Geol. Survey Mon. 20, pp. 48–57, 1902.

[3] King, Clarence, U. S. Geol. Expl. 40th Par. Rept., vol. 1, p. 438, 1878.

boldt formation and are thought by him to be of Pliocene age. The long even sloping ridges of the Humboldt and the larger canyon bottoms are masked by accumulations of recent gravels and wash.

That part of Ruby Valley shown on figure 14 is underlain by fine sands and silts which are in part lake and in part fluviatile deposits.

IGNEOUS ROCKS.

The "granite" is not conspicuous, for it weathers easily and is eroded more rapidly than the sedimentary rocks. It is clearly an intrusive rock, as is shown by the numerous offshoots from the stock and by the contact metamorphism of the adjacent sediments. It was surely formed after the deposition of the Ordovician beds and before that of the late gravels. Its age can not be proved more nearly, but its similarity to igneous rocks at many other places in Nevada whose age is more clearly demonstrated makes it probable that it is late Cretaceous or early Tertiary.

This rock consists of orthoclase, oligoclase-andesine, quartz, microcline, and biotite, with common though not abundant accessory apatite and magnetite. It is a quartz monzonite and even in the center of the stock usually has a porphyritic texture. In the coarser facies the feldspars, usually orthoclase, are the largest phenocrysts, though some quartz grains are of phenocrystic size. The ground-mass is inequigranular and few of the constituent minerals have crystal faces. The marginal facies of the stock are somewhat finer grained than the center, yet as a whole the rocks vary little in texture. The small offshoot dikes have a microgranular ground-mass studded with quartz, orthoclase, and biotite phenocrysts.

Aplite is not abundant, though some small dikes free from dark minerals were noted northeast of Joy. The only basic dike seen was a highly altered dike 10 to 18 inches wide, which the thin section shows to be made up of orthoclase, actinolite, and biotite, with some large quartz phenocrysts.

ALTERATION OF THE ROCKS.

The intrusion of this quartz monzonite porphyry has not caused much metamorphism of the surrounding rocks. The lighter-colored limestones north of Water Canyon have suffered more than the dark dolomitic limestones south of the stock. Where the metamorphism is greatest in the limestone there is a zone of light-greenish lime silicate rock from 20 to 40 feet wide. Some of the shales in the central part of the south fork of Water Canyon have been altered to hornstones.

Along the veins in the quartz monzonite porphyry narrow belts of bleached rock have resulted from the development of calcite and sericite in the feldspars and the alteration of the biotite to masses of chlorite and carbonates. Most of the quartz and apatite crystals in these altered zones have remained unaltered. At the Oddie tunnel this type of alteration is further advanced than at any other place seen in the district. In a belt about 40 feet wide and of undetermined length the quartz monzonite porphyry is altered to a soft white mass which still retains the texture of the original rock. It consists of sericite, calcite, and kaolinite set with little-altered quartz crystals. Some pyrite is disseminated throughout this body and occasional stains of iron oxide and copper carbonates give it somewhat the appearance of the leached portions of some of the " porphyry copper " deposits.

ORE DEPOSITS.

HISTORY.

So far as could be learned, the earliest discoveries in the Bald Mountain district were the silver deposits about 4 miles southeast of Joy. The State mineralogist of Nevada[1] says that the Nevada claim was located August 20, 1869. He describes the mine as being between two peaks 8 miles south of the Overland Stage Road, in the vicinity of two mineral belts, one of free metal 600 yards wide and 4 miles long east of the south peak, and one of base metal 500 yards wide and 2 miles long near the summit of the south peak. He adds that the ore of the Nevada claim carries iron, antimony, lead, and a trace of copper, besides silver chloride that gives it a value of $128 a ton. Between $16,000 and $20,000 is said to have been taken from the surface works in this deposit.

It seems probable that some of the ore bodies at Water Canyon were known in the early days, but it appears doubtful if they were worked much before 1876. The Copper Basin and the old shaft on the Carbonate group were probably exploited in the late seventies or early eighties.

The camp is still in the prospecting stage, and little real development has been done, owing in part to the great distance to mills or transportation. At the Copper Basin property there has been more concentrated work than at any other place in the district, and it is reported that during 1905 and 1906 some copper carbonate ore was shipped from that ground.

TYPES OF DEPOSITS.

So far as development shows all ore deposits of the Bald Mountain district are closely connected with the intrusion of the quartz monzonite porphyry. Ores occur in veins in the igneous rock, as replace-

---

[1] White, A. F., State Mineralogist of Nevada Third Biennial Rept., for 1869–70, p. 78, [1871].

ments in limestone, and in small bodies associated with the lime silicate contact minerals. So far as could be determined, none of these types are strongly mineralized.

Veins in igneous rocks.—In the main stock east of Joy a large number of small white quartz veins cut the quartz monzonite porphyry, usually parallel to a well-defined nearly vertical sheeting that strikes N. 20° E. Some veins strike N. 60° W. to east-west and dip south. The veins are for the most part frozen to the walls, which have been sericitized for short distances on both sides of the quartz. The metallization of these veins is not strong. The principal metallic mineral is pyrite, some of which is cupriferous, although stibnite and marcasite are commonly present in minor quantities. These minerals are more abundant near the walls of the vein and are also seen in the altered wall rocks adjacent to the vein. The central parts of the veins are composed of white quartz, and vugs lined with druses of clear quartz crystals are common. Gold is said to be the only valuable constituent of the veins. Sulphides are found at the surface, and as the water table is very near the surface there is little hope of finding better-grade ores with depth.

Deposits in limestone.—In the limestone areas several deposits of oxidized copper ores occur, either as replacements along simple fractures which usually trend N. 20° E. or N. 60° W., or in zones of brecciation following the same general courses. Their chief value is in copper, though gold and silver are said to be present in varying amounts. The ores are limonitic but contain chrysocolla, malachite, copper pitch ore, and occasionally cuprite and pyrolusite. So far as known no copper sulphides have been found in any of the workings in deposits of this type. In the massive limestones the replacement rarely extends for more than 8 inches from the fissure along which the solutions moved, but where the ores occur in zones of brecciated limestone mineralization may extend over 30 to 40 feet.

Contact-metamorphic deposits.—On the western side of the main stock some copper carbonates were noted in a very small mass of lime silicate rock. A few pits have been sunk.

PLACERS.

The gravels of Water Canyon are said to be auriferous for about 4 miles west of Joy and have been worked to some extent about half a mile west of the settlement, but no washing was under way in September, 1913. Pay dirt 18 to 24 inches thick is said to rest on bedrock and to have an overburden of 6 to 13 feet of wash, in which a little gold is irregularly distributed. The gold is said to be rather coarse, and nuggets worth from $2.50 to $10 have been found. The present water supply is not sufficient to wash these gravels, but they might be worked by dry processes.

Anna tunnel (No. 1, fig. 14).—The Anna tunnel in the south fork of Water Canyon about a mile east-southeast of Joy is owned by W. A. Smith. It runs N. 20° E. for 60 feet on a fracture in quartz monzonite porphyry, along which there has been some movement. No quartz is shown by this work, but the iron-stained bleached country rock on either side of the fissure is said to carry a little gold.

Carbonate group (No. 2, fig. 14).—The Carbonate group of seven claims on the east side of South Bald Mountain belongs to August Munter and Jacob Mayer, of Joy. The No. 1 tunnel is an irregular incline 100 feet long that reaches a depth of about 30 feet along a vertical fracture that strikes N. 25° E. in fine-grained, nearly black dolomite interbedded with quartzite. The wall rocks from 8 to 12 inches on either side of the fissure have been replaced to a small extent by copper carbonates. The small pockets of ore are irregularly distributed, and the mineralization does not appear to be strong.

The Carbonate claim, an old abandoned patent, is in a saddle which marks a zone of faulting that trends N. 25° E. The old shaft is caved at the mouth. What ore remains on the dump is a mixture of limonite and copper carbonate. The zone of brecciated limestone 50 feet wide is all more or less iron stained and shows copper carbonates at several pits. Northeast of the shaft some obscure croppings of quartz monzonite porphyry appear on the line of this fault.

Copper Basin group (No. 3, fig. 14).—The Copper Basin group of 25 claims covers the divide at the head of the south fork of Water Canyon. It is the property of Simonson & Hannon, of Skelton, Nev., but is known as the Scaggs property. The development, which was inaccessible at the time of visit, consists of two crosscut tunnels driven to the ore zone from the main gulch east of the divide and a large open cut with shaft on the divide. The ore makes along a breccia zone of light-colored limestone and quartzite, which strikes N. 60° W. and is thought to mark the largest fault in the camp. Some of the material on the dump is a very highly altered quartz monzonite porphyry impregnated with copper carbonates. The ore on the dumps consists of limonite, copper pitch ore, chrysocolla, malachite, and pyrolusite, and seems to be entirely a replacement of the brecciated rock. It is said that small shipments of ore were made from this property in 1905–6, which carried better than 4 per cent copper and $11 in gold a ton. On the surface the altered and somewhat-mineralized zone seems to be about 20 feet wide, but it is said that underground good ore was found throughout a width of 40 feet.

Copper King group (No. 4, fig. 14).—The Copper King group of nine claims, owned by Robert Raftize, of Joy, is 1½ miles northwest of the camp. Development work in shallow shafts and tunnels has been done on several claims. One shaft 50 feet deep to water is in a nearly vertical, north-striking 50-foot zone of crushed iron-stained limestone that carries irregular pockets of limonitic gold-copper ore. A small exposure of porphyry is near the shaft, and about one-fourth mile to the northwest there is a dikelike mass of jasperoidal breccia, striking N. 50° E., which in some places is stained with copper carbonates. At the east end of this breccia a north-south dike of much-altered quartz monzonite porphyry carries a little pyrite.

Gold King group.—The Gold King claims (No. 5, fig. 14), 21 in number, controlled by Munter, Mayer, and Ziege, of Joy, Nev., cover the central portion of Water Canyon in the quartz monzonite porphyry area.

A number of shallow workings have been sunk on different small quartz veins. The Gold King No. 1 incline one-half mile east of Joy is 30 feet deep on a vein that strikes N. 29° W. and dips 75° WSW. Two to three inches of white quartz is frozen to the walls, which are sericitized for a few inches on either side. The quartz looks rather barren, though occasionally small bunches of pyrite and marcasite are seen. Some black quartz, lining vugs, proves to be an intergrowth of stibnite and quartz. The owners say that gold tellurides have been determined in this ore. No telluride minerals were noted by the writer, but traces of tellurium were found by qualitative chemical tests in some of the material collected by him.

The Essex tunnel, just across the gulch from the Gold King incline, runs S. 20° E. for 146 feet along a series of subparallel quartz veins 4 to 6 inches wide. Two rather persistent quartz stringers about 18 inches apart are the most mineralized. The quartz monzonite porphyry between them is sericitized and contains some disseminated pyrite and is said to assay $19 a ton in gold. Postmineral movement along the hanging wall of this vein is evidenced by grooves which dip 50° to 65° S.

A tunnel one-half mile northeast of Joy runs due east for 160 feet through quartz monzonite along two slip planes, which dip steeply south and are a fraction of an inch to 2 feet apart. The wall rock is slightly altered along them, and a little pyrite is seen in the softened bleached quartz monzonite. On the hill east of the tunnel mouth there are a number of white quartz veins that strike N. 20° E.

A mile northwest of Joy a short tunnel follows a vein that strikes N. 40° W. and dips 85° SW. and is 10 to 16 inches wide. Stibnite and pyrite are sparingly distributed in the otherwise barren white quartz.

The western claims of the Copper King group are on the contact of the limestone and the quartzite, which is marked by a zone of light greenish-gray lime silicate rocks. Some jasperoid has been developed outside of the contact zone, and a few small irregular pockets of copper carbonate ores are present within it.

Mountain View group (*No. 6, fig. 14*).—The Mountain View group west of the Copper Basin consists of seven claims owned by J. W. West, of Joy. A number of prospect pits have opened small bodies of copper carbonate ore near fractures in a dark-blue crystalline dolomite. At the time of visit, Mr. West was starting to sink a shaft which was to be equipped with a whim.

Oddie tunnel (*No. 7, fig. 14*).—The Oddie tunnel is the principal development on the Blue Bell group of 20 claims belonging to August Munter, Jacob Mayer, and Max Ziege, of Joy. It is in the south fork of Water Canyon, a little over a mile east-southeast of Joy. The tunnel runs N. 61° E. for 120 feet through iron-stained sericitized quartz monzonite porphyry. Its last 40 feet is in a zone of intensely altered rock in which there is a minor amount of disseminated pyrite and some small barren quartz veinlets. This mass of highly sericitized and calcitized quartz monzonite porphyry is approximately 40 feet wide on the surface. So far as noted underground it does not carry copper minerals, but on the surface at the east side of the zone a little copper carbonate ore is shown in some shallow workings. It may represent the leached croppings of a mineralized mass of porphyry.

Redbird group (*No. 8, fig. 14*).—The Redbird group of six claims, the property of J. G. Merritt, is in Water Canyon, one-fourth mile west of Joy. A 40-foot breccia of limestone and white quartzite, heavily iron stained and in places containing small irregular pockets of copper carbonate ore, strikes N. 40° W. and dips 20° SW. Numerous open cuts, shallow shafts, and tunnels have been driven into the mass over a distance of 300 feet. The bottom of one shaft 25 feet deep has reached what seems to be very much altered monzonite porphyry. .

Crown Point mine.—The Crown Point mine, about 4 miles southeast of South Bald Mountain, was not visited. It is said to have been worked in 1876. The rich silver ores are said to occur in small pockets irregularly distributed through a width of 20 feet in the vicinity of a vertical fissure trending N. 60° W. and cutting fossiliferous limestone. Parallel to this vein there is said to be a porphyry dike which in places forms the southwest wall of the ore. Stibnite and gray copper are present in some of the ore seen at Joy, though most of it is a soft copper-stained material carrying silver, probably in the form of silver chloride.

Other prospects.—Mr. Albert Dees has a group of six claims about 4 miles north of Joy on the northwest side of Bald Mountain. Some good stibnite ore reported to have come from this property was seen in the settlement. It is said that the ores occur in rather small, irregular pockets in limestone and that little development has as yet been done on any of the claims.

A group of eight claims on the south side of South Bald Mountain, controlled by G. Brant and Max Arnold, of Hilton, Nev., is reported to have small bodies of copper carbonate ore irregularly distributed along fissures in the dark dolomitic limestone.

CHERRY CREEK (EGAN CANYON) DISTRICT.

LOCATION AND ACCESSIBILITY.

The Cherry Creek (Egan Canyon) district (No. 23, fig. 1, p. 18) is in the Egan Range, 90 miles south of Cobre, on the Southern Pacific, and 50 miles north of Ely, the seat of White Pine County. The earliest discovery of mineral in this district was made in Egan Canyon (see Pl. I), a deep cleft that passes entirely across the range and that formed part of the Overland Stage Route between Salt Lake and San Francisco. The Nevada Northern Railway, in Steptoe Valley, which connects Ely with the Southern Pacific, passes about 4 miles east of the town of Cherry Creek, at the mouth of Cherry Creek, a small canyon 5 miles north of Egan Canyon. In the early days Cherry Creek was, next to Hamilton, the largest town in White Pine County, but of late years it has dwindled in size and population. It is still the main supply point of the northern part of White Pine County and the point from which the mail stages leave for Spring Valley east of the Schell Creek Range. (See Pl. I.)

TOPOGRAPHY.

The town of Cherry Creek is at the upper edge of a long gentle slope that rises about 600 feet in the 4 miles west from the railroad. (See fig. 15.) North and west of the town the mountains rise abruptly by steep slopes marked by many cliffs to an elevation of over 8,000 feet in less than 3 miles. From Cherry Canyon south to Egan Canyon the ridge is low and the hills have rounded slopes rather than abrupt rises. Egan Canyon is a deep narrow notch through the range, south of which the mountains rise abruptly to heights considerably greater than those to the north. West of Egan Canyon a narrow north-south valley is bounded on the west by a northward-trending range of low rolling hills.

GEOLOGY.

The eastern side of the Egan Range, in the vicinity of Cherry Creek and Egan Canyon, is composed of quartzites and argillites overlain by limestone and intruded by small dikes and masses of quartz monzonite.

The older sedimentary rocks are exposed along the east base of the range and are particularly well exposed at Egan Canyon and north of Cherry Creek. They are brownish quartzites in distinct massive beds, which strike north-northeast and dip west at medium angles. Some dark arenaceous, micaceous, and argillaceous shales are interbedded with the quartzites, and near the top of the series the quartzites become thin bedded and nearly white.

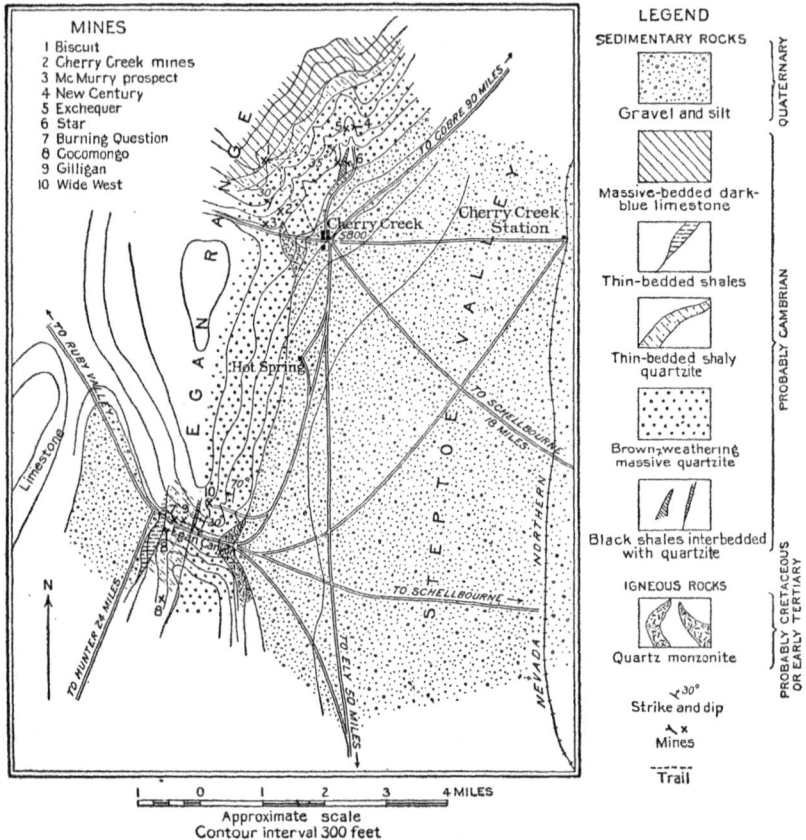

FIGURE 15.—Sketch map of the Cherry Creek (Egan Canyon) district, White Pine County, Nev.

Emmons[1] considered these sediments to be older than Carboniferous, and Spurr[2] called them Cambrian.

The quartzites and shaly quartzites are at least 3,000 feet thick. They are thought to be the equivalent of the Cambrian Prospect Mountain quartzite of the Eureka section, described by Hague.[3]

---

[1] Emmons, S. F., Egan Canyon district: U. S. Geol. Expl. 40th Par. Rept., vol. 3, pp. 445–449, 1870.

[2] Spurr, J. E., Descriptive geology of Nevada south of the fortieth parallel and adjacent portions of California: U. S. Geol. Survey Bull. 208, pp. 47–49, 1903.

[3] Hague, Arnold, Geology of the Eureka district, Nev.: U. S. Geol. Survey Mon. 20, pp. 34–36, 1892.

At the west end of Egan Canyon thin-bedded green and red, somewhat calcareous shales overlie the quartzites. Northwest of Cherry Creek (see fig. 15) a blue-gray crystalline limestone overlies the quartzite, with dark shales near the base of the limestone. These beds are probably to be correlated with the Cambrian Eldorado limestone of the Eureka section. West of the area examined by the writer Silurian, Devonian, and Carboniferous rocks were identified by Spurr.[1]

On the east side of the range the sediments strike N. 10°–30° E. and appear to have a monoclinal dip of 30°–45° W. Spurr[2] says that on the west side of the mountains the sedimentary rocks dip east and that the range is cut from a shallow syncline whose axis strikes north-northeast.

IGNEOUS ROCKS.

Quartz monzonite.—Small masses of quartz monzonite porphyry are exposed at the mouth of Egan Canyon and in Cherry Canyon about a mile west of the town. A narrow dike of this rock outcrops west of the Star mine and is exposed in the west drifts on the seventh and eighth levels. The quartz monzonite is a porphyritic rock with pink orthoclase phenocrysts up to one-half inch in largest dimension. The fresh rock is light gray, but most of the exposures are weathered to a light tawny gray. When examined under the microscope the rock is seen to consist of feldspar, quartz, dark greenish-brown biotite, and a little brown hornblende. Orthoclase and oligoclase-andesine are in equal amounts. Apatite, magnetite, and titanite are common though not abundant accessory minerals. The quartz monzonite porphyry dike on the sixth and seventh levels of the Star mine is almost completely altered to sericite and calcite, though the porphyritic texture is well preserved.

Diabase.—A specimen of what was supposed to be biotite schist, obtained on the tunnel level of the Star mine about 800 feet west of the shaft, proved to be a fine-grained diabase consisting of plagioclase, clove-brown augite, biotite, and a little magnetite. The rock is nearly black and in hand specimens resembles the black arenaceous shaly quartzites which lie on both sides of the dike. The dike seems to be older than the vein, which cuts across all the formations on this level.

ORE DEPOSITS.

HISTORY AND PRODUCTION.

The first discovery of ore in this region was made in 1861 on the Gilligan vein, in Egan Canyon, by John O'Dougherty,[3] one of a

[1] Spurr, J. E., Descriptive geology of Nevada and adjacent portions of California : U. S. Geol. Survey Bull. 208, pp. 48–52, 1903.

[2] Idem, p. 53.

[3] Browne, J. R., Mineral resources of the States and Territories west of the Rocky Mountains for 1867, p. 411, 1868.

party of immigrants who were following the Overland Stage Route to California. He built the first mill in Nevada in 1864 to treat the ores of the mine. Stretch[1] reports that the total production of the mine up to 1866 was $60,000.

The records of the Gold Canyon Mining district begin in 1863, but the first entries are of bills of sale of previously located claims. In 1874 and 1875 Gen. Rosecrans was operating the Gilligan vein for the San José Mining Co. The Cherry Creek district was cut off from the Gold Canyon district in 1872, and the boundaries of this district, as given in the first volume of the district records, are as follows:

Beginning at the mouth of Cherry Canyon, running north along the east base of Egan Range to the mouth of Goshute Creek, thence west through Goshute Canyon to west base of mountains, thence south to the head of Cherry Creek, thence east to point of beginning.

The first claim recorded in the Cherry Creek district was the Teacup, located September 21, 1872, by John Corning and Peter Carpenter.

It is said that the mines of this district were most actively worked between 1872 and 1883, during which time the population of Cherry Creek was about 6,000. There was also a fair-sized town at the head of Egan Canyon up to 1877. In 1884 mining began to wane, and in 1893, with the demonetization of silver, it practically ceased. Since 1895 some work has been done at various properties by different companies, particularly at the Star, Exchequer, and Biscuit mines.

The production of the early days is not known, though estimates range from $6,000,000 to as high as $20,000,000. The following table shows the production from 1902 to 1912 inclusive, as reported to the United States Geological Survey.

Production of the Cherry Creek district, White Pine County, Nev., 1902–1912, inclusive.

Year.	Gold.	Silver.	Copper.	Lead.	Total value.
	Fine ounces.	*Fine ounces.*	*Pounds.*	*Pounds.*	
1902	506.87	14,157	....	....	$17,838
1903	5,170.26	33,493	....	....	120,947
1904	524.44	5,325	....	....	13,950
1905	146.43	15	....	....	3,036
1906	241.88	52	....	....	5,035
1907	134.77	4,779	....	....	5,940
1908	78.55	2,870	....	....	3,045
1909	325.27	20,125	315	56,744	19,670
1910	1,451.01	88,247	9,622	121,759	84,227
1911	73.19	2,287	410	10,138	3,232
1912	11.03	572	1,151	35,575	2,371
	8,663.70	171,922	11,498	224,216	279,291

[1] Stretch, R. H., Nevada State Mineralogist Ann. Rept., 1866, p. 102, 1867.

There are two somewhat overlapping types of veins in the Cherry Creek (Egan Canyon) district. One type, represented by the Wide West, Cocomongo, and possibly the McMurry prospect, has its principal value in gold, carried in a white quartz gangue that shows a minor amount of pyrite and less galena. The other and by far the most important in the district is represented by the Exchequer, Star, Biscuit, Cherry Creek, and Gilligan veins, which carry galena, sphalerite, pyrite, and rich secondary silver minerals. That there may be a transition between these two types is strongly suggested by the Exchequer-New Century vein, in Exchequer Canyon, north of Cherry Creek. West of the canyon, on the Exchequer ground, the quartz carries the base-metal sulphides and contains more silver than gold; and east of the canyon, on the New Century ground, what appears to be the same vein is not strongly mineralized, but carries pyrite, gold, and silver. It may be that the gold and silver bearing portions of the veins are those parts which have not been strongly mineralized, and that the ore shoots will all prove to be of the lead-zinc type, carrying more silver than gold. In both types the veins are strong in the quartzites but tend to finger out where the fissure enters argillaceous shales. This is particularly well shown on the Gilligan and Star veins.

At the upper tunnel on the Cherry Creek Co.'s property the ore body is a mineralized quartz monzonite dike carrying galena, sphalerite, and pyrite. It is evident that the dike was metallized after its consolidation, and it is believed that most of the veins of the district were formed shortly after the intrusion of the quartz monzonite.

The lead-zinc veins have been crushed since the deposition of the original ores and have been enriched by descending waters which have deposited rich silver minerals such as argentite, proustite, and an antimonial silver-lead-copper mineral of uncertain composition. The enrichment of the Star vein, as indicated by the largest stopes, was greatest between the third and seventh levels but extended to a depth of 600 feet vertically below the croppings. At the Biscuit mine argentite and copper carbonates are said to have been found to a depth of 1,100 feet on the dip of the vein.

THE MINES.

CHERRY CREEK.

Biscuit mine (No. 1, fig. 15).—The Biscuit mine is about 2½ miles northwest of Cherry Creek, near the summit of the range at an elevation of 7,800 feet. This property, known as the Teacup mine, was worked in the early days and again in 1907 and 1912. It is said to have produced about $3,000,000 worth of silver. It is developed through an incline shaft equipped with a 60-horsepower gasoline hoist. The incline is 1,100 feet deep, dipping 30° W. to the 60-foot level, 50° to the 100-foot level, 40° to the 800-foot level, and about 10° below that level. It follows under a 25-foot bed of greenish gray shale, interbedded in limestones, which strike N. 10° E. Some movement has occurred along this contact. Lenslike bodies of white quartz are found in the limestone under the gouge, the largest lens being between the 100 and 200 foot levels south of the shaft. There are many drifts at many levels in the shaft, but none of them appear to be long, and most of them were caved in October, 1913. Much of the white quartz of the ore is barren looking but in certain places is speckled with minute particles of a soft black metallic mineral containing silver and a little copper and sulphur. This is either argentite or stromeyerite, but in the few specimens found there is not enough of the mineral to absolutely determine the composition. A small amount of copper carbonate has been deposited on joints of the richer ore.

Cherry Creek mines (No. 2, fig. 15).—The Cherry Creek Mines Co. owns 22 claims on the hills north of Cherry Canyon, about 1½ miles west of town. A patent notice at the lower tunnels places the claims in secs. 25 and 36, of T. 24 N., R. 62 E. The main ore body is opened by a tunnel 300 feet above the canyon bottom, on the north hill. This tunnel, caved at the mouth, is said to be 600 feet long on the lode, which strikes N. 30° E. and dips steeply west. The lode is in quartzite about 100 feet west of and parallel to a large dike of quartz monzonite. It is said to average 12 feet in width and to open out in places to a maximum of 30 feet. The ore on the dump is a highly altered porphyry containing abundant pyrite and galena and some sphalerite irregularly scattered through the altered rock. Some pieces of nearly pure sulphides indicate that the metallic minerals were concentrated in places. In some of the more disseminated ore the crystals of pyrite are coated with thin black films of chalcocite.

The National tunnel, whose portal is on the north side of Cherry Creek, about 1¼ miles west of town, has been driven N. 10° W. for 1,400 feet but has not as yet cut the ore body exposed in the upper work. The tunnel is started near the west side of a dike of quartz monzonite 250 feet wide that strikes N. 31° E. and dips 30° W. In the tunnel the igneous rock is seen to be cut by a series of eastward-trending fractures that dip 20°–30° N.

In the Brady tunnel, whose mouth is a little lower than and west of the National tunnel, some small quartz veins in the quartz monzonite strike N. 80° W., dip 30°–50° S., and carry a little pyrite. The wall rock is sericitized for a short distance on each side of these small veins.

McMurry prospect (No. 3, fig. 15).—The McMurry prospect is on the north side of Cherry Canyon, 2 miles west of town. An open cut exposes a zone of crushed, iron-stained, silicified quartzite 15 feet wide, which strikes N. 35° E. and dips 70° W. Some narrow quartz stringers cut the zone, and along them the quartzite contains minute specks of pyrite and molybdenite. The value of this ore is not known, but it may carry a little gold.

New Century tunnel (No. 4, fig. 15).—The New Century tunnel is driven to develop the Old Imperial, Ben Butler, Jim Blaine, New Century, and Emma claims, on the ridge east of Exchequer Canyon, about 2 miles north of Cherry Creek. The main workings in Exchequer Canyon consist of a drift tunnel 600 feet long, which cuts a vertical shaft 100 feet from the mouth. The shaft is 275 feet deep but was not accessible. The tunnel follows a quartzite fault breccia, 2 to 6 feet in width, that strikes N. 70° E. and stands nearly vertical or dips steeply north. This breccia constitutes the ore. At the end of the drift the vein is cut by a fault that strikes N. 10° W. and dips 60° E. Below this fault there is 10 feet of soft altered quartzite that is said to carry some silver.

The quartzite fault breccia has been partly cemented by quartz, and the healed rock shows very small crystals of pyrite. In some joints in this breccia there are small patches of copper silicate stain and here and there a little yellow stain that may be lead carbonate. No large bodies of sulphides have been found, even at the bottom of the shaft, but the ore is said to average about 30 ounces silver and a little gold to the ton. One piece of ore found on a dump, said to have come from the 250-foot level, contains minute specks of galena and pyrite, and under the blowpipe shows iron, lead, silver, copper, and a very faint black antimonial mirror in the closed tube, thus indicating the probable presence of gray copper.

Exchequer tunnel (No. 5, fig. 15).—The Exchequer tunnel runs west from the canyon on the same vein as the New Century. A vertical shaft at the mouth of the tunnel, said to be 186 feet deep but containing 100 feet of water, is equipped with cage, 30-horsepower steam hoist, and 6-drill compressor. The tunnel is said to be 800 feet long, but the open stopes near the mouth can not be crossed. The ore shoot is said to have extended for 600 feet along the tunnel level and to have been 2 to 3 feet wide, expanded to 8 feet in places. Near the shaft the vein splits, but reunites 100 feet west. The south part of the vein was the most productive. The ore above water level is very siliceous and looks like that at the New Century. It is said that much of the rich surface ore mined in the early days contained considerable horn silver. Some ore found on the dump, said to have come from below water level, is base, consisting largely of galena, sphalerite, and pyrite. In one specimen of this sulphide ore a small vug lined with minute quartz crystals contained also some thin scales of a dark metallic mineral that looks like dark ruby silver but was in too small a quantity to allow definite blow-pipe tests. It is said that the Exchequer vein, which has yielded about $3,000,000 worth of silver, was to be reopened in the fall of 1913.

Star mine (*No. 6, fig. 15*).—The Star mine, consisting of four claims on the west side of the mouth of Exchequer Canyon, is about 1½ miles north-northeast of Cherry Creek and is the property of the Glasgow Western Exploration Co., whose business was in the hands of Joseph Ralph in March, 1913. The mine is developed by a vertical shaft approximately 750 feet deep, with eight levels, and by a tunnel about 1,200 feet long, which intersects the shaft 340 feet below the collar. The old workings above the tunnel level are caved and much of the work below that level is inaccessible. Below the tunnel level two subparallel veins about 30 feet apart are exposed in the workings. On the fourth level the drifts on both veins extend about 400 feet west and 600 feet east of the shaft; on the fifth level they extend at least 200 feet east and 800 feet west; on the sixth level 1,000 feet west; and on the seventh level 1,200 feet west. The east drifts on both the sixth and seventh levels are inaccessible but are not very long.

The north vein has an average strike of N. 78° W., and the south vein strikes N. 72° W. Their intersection is shown in both the sixth and seventh levels, about 1,000 feet west of the shaft. At some places the veins are vertical, but at most places along the drifts they dip steeply south, cutting directly across the bedding of quartzite and shales, which strike N. 30° E. and dip 35° NW. Both veins are strong in the quartzites, even where they are thin bedded, but finger out in the shales, as is clearly shown at the west end of the tunnel level, at the east end of the fourth level, and at the face of the west drift on the seventh level.

The veins are cut by a number of minor faults that strike north or N. 30° E. and stand vertical or dip steeply west. The movement along these breaks seems to have been normal, for the vein west of the faults is displaced 2 to 4 feet to the south. A fault on the fourth level about 250 feet east of the shaft strikes N. 10° E. and dips 45° E. The fault zone is 20 feet wide and is marked by a very thick gouge and considerable breccia. East of the fault zone the vein is displaced 20 feet north. The movement along this fault is believed to have caused the repetition of the shale seen on the surface east of the shaft and in the shaft to a depth of 160 feet. A second important fault is exposed at the west ends of the drifts on the sixth and seventh levels. On the sixth level the fault strikes N. 70° W. and dips 76° W., but on the seventh level it strikes N. 40° W. and dips 50° NE. The movement along the fault as shown at both levels was horizontal. West of the fault a narrow dike of very much altered porphyry, impregnated with pyrite, was probably a quartz monzonite, as phenocrysts of quartz and both orthoclase and plagioclase can be recognized. The feldspars are largely altered to calcite and sericite and these minerals form the groundmass of the rock.

The stopes indicate that the south vein was the most productive. One shoot on this vein, extending for 900 feet along the drift, varied from 1 to 6 feet and averaged about 2 feet in width. It lies below argillaceous shales, exposed at the west end of the levels. The vein fingers out and is not mineralized in the shales. The western end of the shoot pitched 45° W. to the 400-foot level and 25° W. between the fourth and eighth levels. The ore consists of galena, sphalerite, and pyrite, carried in a white quartz gangue. The vein shows depositional banding but has been crushed by postmineral movement. Rich secondary silver minerals are seen in small amounts in cracks of the original ore, even at the eighth level, which is at least 400 feet below the original water level. Ruby silver and a light-gray mineral, probably tetrahedrite, that contains silver and lead as well as copper seem to be the most valuable secondary minerals, though secondary chalcopyrite is more abundant. It is said that polybasite and argentite have been found in ore from the eighth level. In one specimen of enriched ore the silvery-gray mineral, together with chalcopyrite, replaces the original galena but occurs in veinlets in the sphalerite. The ore in this shoot on the eighth level is said to average 0.32 ounce gold and 17.47 ounces silver a ton. Some of the enriched ore carries as much as 614 ounces of silver a ton.

The north vein is narrower than the south vein, averaging about 1 foot wide. It is composed of quartz and pyrite, with much less sphalerite and galena than the south vein. The ore is said to average 0.24 ounce gold and 6.47 ounces silver a ton.

The Star shaft is vertical and below the tunnel level is north of both veins. It is well timbered and is said to be in heavy ground. The surface equipment is a large double cylinder steam hoist and six-drill compressor. The mine makes a great deal of water. In 1913 five men were employed in pumping, and it was said that the mine would fill to the tunnel level in a few months if pumping were stopped. A partly dismantled mill stands at the mouth of the tunnel, to which level all ore was hoisted. The mill has two Blake crushers, twenty 1,200-pound stamps, and Wilfley tables and Frue vanners, with the necessary classifying apparatus. For a time the tailings and middlings were reground in a tube mill and cyanided, but in 1913 the leaching tanks had been removed. The mill is run by a 150-horsepower gas engine supplied by a 200-horsepower gas producer.

EGAN CANYON.

Burning Question group (No. 7, fig. 15).—The Burning Question claims are on the north side of Egan Canyon near its western end and southwest of the Gilligan vein. Mr. G. O. Kellsey is developing a 20-foot zone of quartzite, sparingly impregnated with pyrite, which is said to carry some gold. It was reported in the mining journals in the spring of 1914 that he had opened a pocket of rich gold ore. This quartzite is about 100 feet east of the shale belt that cuts across the head of Egan Canyon.

Cocomongo claims (No. 8, fig. 15).—The Cocomongo claims extend for about 3 miles along the west side of the Egan Range south of Egan Canyon. At a number of places narrow quartz veins cutting the quartzite strike about N. 15° E. and dip 30° E. or trend eastward and stand vertical. The northward-trending lodes are about parallel to the strike of the quartzites but dip east across the westward dip of the sediments. None of these veins are well mineralized. They consist of white quartz, in some places iron stained and in a few places containing a little pyrite. It is said that one or two pockets of very rich free gold ore have been found at the surface. The veins apparently do not persist in the shales which overlie the quartzite, and they appear to finger out in a narrow belt of shale interbedded with the quartzite.

The principal development work is a 600-foot crosscut tunnel about 2 miles south of Egan Canyon. On the south side of the head of Egan Canyon an irregular tunnel about 400 feet long exposes a few narrow quartz veinlets and a series of northward-trending faults that cut off the veins. A 40-ton mill on this property, equipped with 1,000-pound Nissen stamps, plates, and concentrating tables, all driven by a 50-horsepower gasoline engine, has apparently never been used.

Gilligan vein (No. 9, fig. 15).—The Gilligan vein, on the north side of Egan Canyon, near its western end, was the first vein discovered in the district. It was worked by the Social & Steptoe Co. for many years and in 1869 was developed to a depth of 400 feet.[1] In 1913 the incline was said to be 900 feet deep and the drainage tunnel, which connects with the 400-foot level, 1,200 feet long. The lower workings and the tunnel were under water in October, 1913.

[1] Emmons, S. F., Egan Canyon district: U. S. Geol. Expl. 40th Par. Rept., vol. 3, pp. 445–449, 1870.

The vein strikes N. 70° E. and dips 65°–68° N., cutting quartzites and shales. In the quartzite the vein ranges from 2 to 8 feet in width and has a beautifully marked hanging wall. The footwall is not well defined, and quartz stringers are abundant in it for some distance below the vein. At the 100-foot level the vein enters a shale belt 365 feet east of the shaft. In the shales, which strike N. 30° E. and dip 30° W. the vein is represented by 6 inches to 2 feet of crushed shale with almost no quartz and showing no mineral. In the east drift, 265 feet from the shaft, the vein again enters quartzite and is well marked, being about 6 feet wide between good walls. The filling is crushed quartzite, quartz, and some sulphides, but it was apparently not rich enough to pay for mining. Most of the ore has been removed from the parts of the mine which could be entered, but a few pillars show that the vein consisted of white quartz and sulphides that have been crushed since the mineralization. The quartz next the hanging wall appears to carry less mineral than that 8 to 10 inches above the footwall, the two streaks being separated by a thin selvage. Galena appears to be the most abundant sulphide in the ore that is left, though some soft black argentite was seen on a specimen found on the dump. Thin coatings of lead and copper carbonates are seen on joints in the ore at the 100-foot level, and horn silver is said to have been the most valuable mineral in the ore from a surface stope northeast of the main shaft. Emmons[1] reported that the ore ran from $40 to $200 a ton and that the bullion was two-fifths gold and three-fifths silver.

The mill in Egan Canyon, above the tunnel, belonging to this company, was in bad repair in 1913.

Wide West group (No. 10, fig. 15).—The Wide West group of twelve claims and four mill sites is in the hills north of Egan Canyon on the east side of the Egan Range. A 5-stamp amalgamation and concentration mill is at the mouth of Egan Canyon. The Wide West vein, in the head of Gold Canyon, a small steep gulch about one-fourth mile north of Egan Canyon, is developed by two tunnels. The upper tunnel runs north parallel to the bedding of thin-bedded quartzite (which dips 70° W.) for 172 feet, to the vein, which strikes N. 60°–70° E. and dips 32°–50° S. and is followed by a drift for 220 feet. A winze connects this tunnel with the lower one, which runs N. 35° W. for 380 feet to the vein, cutting the bedding of the quartzites at an acute angle. At this level the drift N. 50° E. on the vein is 125 feet long and the drift southwest is 300 feet long. The vein as exposed by these tunnels ranges from 2 inches to 2 feet in width (average, 10 inches). It is frozen to the hanging wall but is separated from the footwall by a thin clay selvage. It consists of

---

[1] Emmons, S. F., op. cit., p. 447.

barren-looking milky-white quartz that is only slightly stained with iron oxide on the joints and in places shows a few widely scattered crystals of pyrite. The pyrite at the upper tunnel level is largely altered to a dark-brown oxide. The ore at the lower tunnel level is less oxidized and shows a few specks of galena in its more heavily mineralized parts. This ore is said to carry some gold and silver, but from all that could be learned is rather low grade. The matrix of the quartzite near the vein is somewhat sericitized and shows some minute crystals of pyrite between the quartz grains which form most of the rock.

HUNTER DISTRICT.

LOCATION AND ACCESSIBILITY.

The Hunter district (No. 26, fig. 1, p. 18) is on the west side of the Egan Range about 25 miles south-southwest of Cherry Creek, which is its shipping and supply point. The Hunter mine is about 10 miles northwest of Steptoe post office, which is on the west side of Steptoe Valley, 23 miles north of Ely. (See Pl. I.)

GEOLOGY.

SEDIMENTARY ROCKS.

In the vicinity of the Hunter mines the mountains are formed of dark massive-bedded dolomitic limestones which dip 25°–30° W. and overlie the Cambrian quartzites exposed farther north in the range. Among some fossils collected by the writer at the Hunter mine Edwin Kirk determined Bryozoa and *Cyathophyllum* sp., which he states occur in the Nevada limestone, of Devonian age, of the Eureka district.

IGNEOUS ROCKS.

The limestones at the mines are intruded by at least three large dikes of granite porphyry. The igneous rocks are altered and exact determination of their constituents is not possible. They contain conspicuous rounded, somewhat smoky quartz phenocrysts one-eighth inch in maximum diameter; other phenocrysts that appear to have been feldspars but that are now altered to a soft white substance, and, generally, a few flakes of dark greenish-brown biotite. Examined in thin sections the phenocrysts that were thought to be feldspars are seen to consist of a mixture of quartz, an isotropic substance with index higher than balsam, and some calcite. The groundmass of the rock is microgranular, consisting of quartz, altered feldspar, and some biotite. It exhibits less alteration than would be expected from the complete breaking down of the feldspar phenocrysts.

ORE DEPOSITS.

HISTORY AND PRODUCTION.

The lead-silver ores of the Hunter mines were discovered in 1871, according to Whitehill,[1] and were most extensively worked between 1877 and 1884. At that time the ores smelted on the ground were said to carry about 48 ounces silver and 48 per cent lead a ton. Some very rich silver chloride ore is said to have been mined at that time in a pipelike deposit on the Crown Point lode. The 18 patented claims in this district were bought by the Vulcan Mining, Smelting & Refining Co. in 1907. In 1913 this company, under the direction of Mr. H. Ornauer, was reopening the main workings on the Vulcan and Copperhead Split ore bodies.

According to data collected by the United States Geological Survey, the total production of the Hunter properties from 1905 to 1908, inclusive, was 28,072 ounces silver, 118,584 pounds of copper, and 412,305 pounds of lead, having a total value of $60,316.

DEVELOPMENT.

The main development at the Hunter district is through a crosscut tunnel about 1,700 feet long, which trends southeast into the mountains. This tunnel intersects the Copperhead Split and the Vulcan ore bodies, the latter at a depth of 400 feet on the dip. The Vulcan is also opened higher up by a 300-foot drift tunnel, and the ore shoot, which had a drift length of 75 feet, is said to be stoped to the lower tunnel level. The ore bins and power house are at the mouth of the main tunnel. A steam-driven compressor supplies air for the hoist at the winze on the Copperhead Split ore body and for two hammer drills.

Water from a spring 3 miles north of the mine is piped to both the power house and camp. Sufficient piñon for mine timbers can be found on the mountain east of the mine.

GEOLOGY OF THE TUNNEL.

The first 400 feet of the tunnel are through somewhat crystalline, rather dark gray fetid-smelling dolomitic limestone in beds 1 to 3 feet thick, which strike about north and dip 30° W. At 400 feet the tunnel cuts the western side of a 75-foot dike of granite porphyry which dips steeply west. The contacts are sharp and show no contact metamorphism along either side. Postintrusion faulting with horizontal movement is shown. The Copperhead Split ore bodies lie along this dike.

A second dike, 80 feet wide, is cut at 800 feet from the portal, and a third, 60 feet wide, at 970 feet. From a point 1,320 feet from the

---

[1] Whitehill, H. R., Nevada State Mineralogist Fourth Biennial Rept., for 1871–72, p. 145 [1873].

portal to the face the tunnel is in granite porphyry. The Vulcan ore body lies west of this mass of porphyry, but the fault along which the ores occur has cut into the dike at places.

The limestones along these dikes appear to be unmetamorphosed even at the contacts, and the dikes themselves do not show any change in texture or size of grain. Postintrusion faulting is seen along all the dikes. Movement along practically all these faults seems to have been horizontal, as evidenced by deeply cut grooves on the walls. Much breccia has been formed along most of the faults and it is in these brecciated masses that the ore bodies occur.

OCCURRENCE AND CHARACTER OF THE ORE BODIES.

The Copperhead Split ore body is cut by the tunnel 575 feet from the mouth, where it strikes N. 30° E. and dips 75° E. It is a limestone fault breccia about 5 feet in width below a dike of altered granite porphyry. The lead carbonate ores occur as irregular masses in the breccia and are usually associated with coarsely crystalline white calcite. In a winze, 130 feet below the tunnel level, the breccia strikes north and dips 45° E. A stope at this level is 150 feet long, 15 feet high, and 6 feet wide. The ore, an ocherous lead carbonate containing some residual masses of galena coated with thin films of anglesite, seems to have been formed by partial replacement of the limestone breccia, as it includes many fragments of limestone. In some places in this stope malachite is rather abundant, and most of the limonitic ore carries copper, even where it shows little copper carbonate. The assay returns of some of the shipments of ore mined in 1913 show the lead ore to carry on the average 0.01 ounce gold, 15 ounces silver, 13 to 16 per cent lead, 4.5 per cent copper, and 5 per cent zinc. Sorted copper carbonate ore shipped in 1913 carried 27 to 34 ounces of silver and 10 to 11 per cent copper.

The Vulcan ore body is cut 1,500 feet from the mouth of the tunnel. It strikes about S. 30° E. and dips irregularly, varying from vertical to 40° NE. or 70° SW. At the tunnel level the ore is in altered granite porphyry, but in the upper workings and in a drift 200 feet north of the tunnel it occurs in a breccia at the contact of the limestones and the dike. As shown in the drift the fault breccia along the contact is 30 feet wide. The ore is all oxidized and is similar to that being stoped from the Copperhead Split body, though it is said to have carried less copper.

GRANITE (STEPTOE) DISTRICT.

LOCATION AND ACCESSIBILITY.

The Steptoe or, as it was formerly called, the Granite district (No. 25, fig. 1, p. 18), includes a portion of the east side of the Egan Range about 12 miles long and 4 miles wide. Steptoe post office, at W. D. Campbell's ranch in the southeast corner of the district, is 5½ miles southwest of Granite siding and 23 miles north-northwest of Ely, on the Nevada Northern Railway. (See Pl. I.) Most of the mines are easily accessible, and roads from the mouths of the various canyons lead toward Granite siding. In recent years the roads have been used to some extent by woodchoppers and are in fair shape. The Cuba mine, about 2 miles west of Steptoe post office, is difficult of access, as many of the workings are on a ridge with precipitous slopes.

GEOLOGY.

This part of the Egan Range is composed of yellow and red weathering quartzites, exposed at the east base of the mountains, overlain by green and red shales, above which lie massive light-gray compact limestones which form all the higher mountains. In the

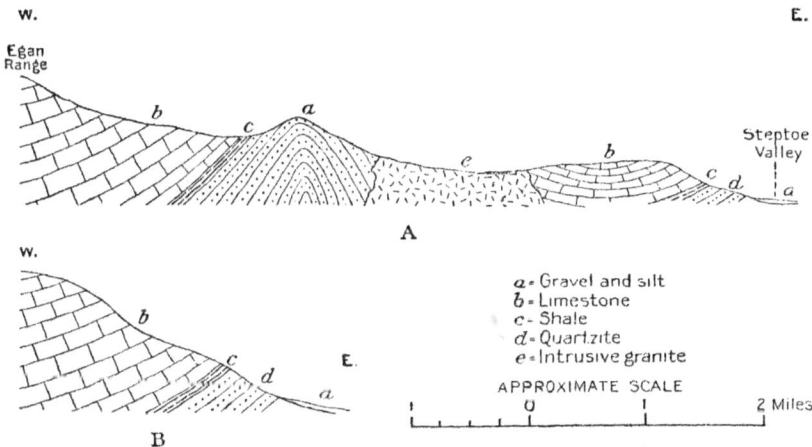

FIGURE 16.—Diagrammatic sections of the east front of the Egan Range, Steptoe district, White Pine County, Nev. A, At Water Canyon, 10 miles north of Steptoe post office; B, 4 miles northwest of Steptoe post office.

northern part of the district, at Water Canyon, the sedimentary rocks are intruded by a large stock of granite that varies from equigranular to porphyritic in texture. In other places in the district there are small dikes and masses of granite porphyry.

SEDIMENTARY ROCKS.

The oldest sedimentary rocks are most extensively exposed in the vicinity of Water Canyon, about 10 miles north of Steptoe post office. They are quartzites of a rather light color on fresh fractures, but weather yellow or red on the surface. About 2 miles above the mouth of Water Canyon they are folded into a tight anticline, across whose axis the stream has cut. At the east they are cut off by a large stock of granite. Northeast of the mouth of Water Canyon limestones overlie the igneous rock. (See fig. 16, A.) In the western limb of this anticline the quartzites are lighter colored and thinner bedded near the top of the series than in the lower part. The quartzites are overlain by about 150 feet of greenish, drab, and red shales, which grade into the overlying limestone. At the base the limestone is dark gray and dense and lies in beds 6 inches to 1 foot thick. The higher beds of the series are more massive, and alternating light and dark gray bands are conspicuous.

About 4 miles northwest of Steptoe post office the limestones form practically all of the mountains, though the underlying shales and thin-bedded light-colored quartzites appear in the foothills. (See fig. 16, B.)

Though no fossils were found in any of these sedimentary rocks during the reconnaissance, the writer has little hesitancy in referring them to the Cambrian. They probably are the equivalents of the Prospect Mountain quartzite and the Eldorado limestone of the Eureka section.

IGNEOUS ROCKS.

About 2 miles above the mouth of Water Canyon, in the northern part of the district, a body of granite intrudes both the quartzite and the limestones. (See fig. 16, A.) This is the south end of what seems to be a large body of granitic rock that extends for some miles along the east front of the range toward Cherry Creek. The typical rock of the middle portion of this stock is a coarse, inequigranular to porphyritic granite, which weathers in rounded forms and produces an immense amount of coarse arkosic sand. At the west side of the stock the rock becomes much finer grained and appears to contain less of the iron-bearing minerals than the typical rock. In the coarse-grained rock large crystals of quartz, pink orthoclase, and biotite are conspicuous and in many places are the phenocrystic minerals that are set in an inequigranular groundmass composed of the same minerals and of what seems to be a plagioclase feldspar. In the fine-grained phase of the rock oligoclase is less abundant than quartz or orthoclase, and a little muscovite and what appears to be bleached biotite are present.

Four and a half miles northwest of Steptoe post office, near the Nubagah claim, there is a dike of granite porphyry which has a purplish-red color. Rounded quartz phenocrysts are thickly studded through a fine-grained matrix of quartz, orthoclase, and glass. Some phenocrysts, having the form of feldspars, are altered to quartz, calcite, and iron oxide.

A few hundred feet west of the Cuba ore body some inconspicuous, weathered croppings of a much-altered yellowish-green granite porphyry have small rounded grains of quartz, sericitized feldspars, and some biotite as their phenocrystic constituents. The groundmass is fine grained and seems to be made up of quartz, feldspar, and chlorite.

In general the rocks of the southern two-thirds of the Granite district dip west at low to medium angles. About a mile south of Water Canyon there seems to be an eastward-trending break in the mountains. South of this fault the quartzites are shifted so that they are seen only in the foothills (see fig. 16, B), but due north of it they dip eastward on the east limb of a tight anticlinal fold. (See fig. 16, B.)

ORE DEPOSITS.

HISTORY AND PRODUCTION.

According to Mr. W. D. Campbell, the Ben Hur vein was located by himself in 1894 and was the earliest location in the Granite district. The Cuba and Stinson ore bodies were discovered in 1902 and the Campbell and Blaine groups in 1907. The latest discovery of ore was made in 1913, on the Nubagah claim, by Baird and Campbell. The gold deposits were worked earliest, and Mr. Campbell states that he has recovered about $15,000 worth of bullion from ores mined from the Ben Hur, Stinson, and Campbell groups. This ore was milled in a 5-stamp water-driven amalgamation mill at Mr. Campbell's ranch. Some lead ores from the Cuba and Bunker Hill and Sullivan mines have been shipped to the Utah smelters. In October, 1913, the Cuba was being worked, under lease, by E. E. Vanderhoff, of Ely, Nev. None of the older properties were being worked, but the Nubagah was being prospected.

According to figures collected by the United States Geological Survey, the total production of the Granite or Steptoe district from 1902 to 1912 inclusive was 454.28 ounces gold, 615 ounces silver, and 114,772 pounds of lead, having a total value of $14,633.

TYPES AND OCCURRENCE OF THE ORE DEPOSITS.

Two distinct types of ores are shown in the Granite district. One type carries gold and a little silver in quartz veins in quartzite. The Blaine, Campbell, Stinson, and Ben Hur, all deposits of this type, have a gangue of white quartz or brecciated quartzite. Metallic minerals are very scarce, and so far as developed all of the gold is free milling. The second type carries lead and a very little silver in fissure replacements in limestone. Carload shipments of sorted galena from the Cuba are said to run 74 to 78 per cent lead and 2.75

ounces silver a ton. The most important metallic mineral is galena, which is found at the surface, with a minor amount of cerusite and anglesite. Coarsely crystalline white calcite is the characteristic gangue mineral, though some siderite has been found in the Cuba ore body.

THE PROPERTIES.

GOLD VEINS.

Alvin mine.—The Alvin vein on the W. D. Campbell group is opened by a 150-foot shaft, a drift tunnel 200 feet long that intersects the shaft at a depth of 75 feet, and a 75-foot drift on the 150-foot level of the shaft. This work is in low foothills about 6 miles northwest of Steptoe post office. The vein cuts thin-bedded light-yellow siliceous limestones and calcareous sandstones that lie at the top of the quartzite series. It strikes N. 50° E., stands nearly vertical, and ranges in width from 8 inches to 4 feet (average about 18 inches). The filling is a yellow clay carrying crushed fragments of limestone and some calcite and quartz. No metallic minerals are visible. Ore from the tunnel level is said to have averaged $8 a ton in gold, and that from the bottom level about $3 a ton. A sample of ore from the tunnel level, panned by the writer, gave a very small concentrate of pyrite and magnetite.

Ben Hur mine.—The Ben Hur vein is opened by a series of open cuts and a drift tunnel 300 feet long, about 3 miles northwest of Steptoe post office. In this vicinity the thin-bedded, light-colored quartzites strike north and dip 60° W. The vein strikes N. 44° E. and dips 53° SE. It is filled with barren-looking white quartz, not much stained with iron, that averages 10 inches in width but in places widens to a maximum of 14 inches. Faulting along the footwall of the vein has crushed the quartz and wall rock so that in places there are 2 feet of yellowish clay with fragments of quartz and quartzite. The hanging wall is very irregular, and the quartz is frozen to it in most places. The ore near the surface is said to have carried about $10 in gold a ton, but at the tunnel level it is said to be too low grade to pay mining and milling charges.

Blaine prospect.—The Blaine property, near the head of Water Canyon, is about 12 miles north-northwest of Steptoe post office. The main development on this property is a tunnel driven southwestward from Water Canyon for 1,200 feet. The first 150 feet of the tunnel trends S. 50° W. along a 2 to 4 inch vein of white quartz that dips 60° SE. This vein cuts quartzite, but postmineral movement along both walls has produced a narrow selvage between the quartzite and vein. A fault that trends north and dips 75° W. cuts off the vein 150 feet from the portal. Beyond this vein the tunnel continues as an irregular crosscut in quartzites, in some places following the bedding and in other places following fractures with north or northeast strike. The last 210 feet of the tunnel follows a 2 to 8 inch lode of gouge, crushed quartzite, and vein quartz that strikes N. 60° E. and dips 50° N. The lode is slightly iron stained and contains a few minute crystals of pyrite and a little magnetite. It is said to carry gold, but no free gold was obtained in pannings made by the writer. Grooves on the walls of this vein are horizontal. A 600-foot incline raise on the Blaine vein connects with the surface work, where the vein is seen to be on the west limb of an anticline, in quartzite, about 50 feet east of the lowest shale bed of the overlying series. (See fig. 16, A.)

Stinson mine.—The Stinson veins, the property of W. D. Campbell, about 4½ miles northwest of Steptoe post office, are developed by several short tunnels, open cuts, and a lower tunnel 300 feet long. One 10-inch lode striking N. 40° E. and dipping 35° NW. consists of three subparallel slips, with intervening fractured quartzite. The other lode strikes N. 40° E., dips 35° SE., and averages 10 inches wide. Both fissures are filled with yellowish clay and quartz. Pannings of the ore from these veins give more concentrates than ore from any of the other veins in the district. The concentrates consist of magnetite and a few specks of pyrite. Sorted ore from the surface of these veins is said to have carried from $30 to $80 a ton in gold.

LEAD DEPOSITS.

Bunker Hill and Sullivan mine.—The Bunker Hill and Sullivan ore body, the property of people from Moscow, Idaho, is 8 miles north-northwest of Steptoe post office in light buff-gray, thin-bedded, somewhat siliceous limestones that strike north and dip 40° W. Slightly oxidized galena occurs in irregular masses in a fault breccia 2 to 10 feet wide that strikes N. 30° E. and dips 60° W., and also in small tabular bodies parallel to the bedding of the limestone. The property is developed by a 75-foot crosscut tunnel that intersects a whim shaft about 30 feet below the collar. The shaft is 160 feet deep, but shows little ore below the tunnel level. The galena is clearly a replacement of the limestone. Postmineral movement has crushed both the limestone and ore, producing what is known as steel galena.

Cuba mine.—The Cuba fissure, about 2 miles west of Steptoe post office, strikes N. 40°–43° E. and dips 40°–65° SE., cutting across the bedding of light and dark colored dense limestones in beds 2 to 4 feet thick that strike N. 20° E. and dip 30° W. The fissure has been opened by a number of short tunnels and open cuts for about 3½ miles through a vertical distance of 500 feet. In most places the fissure is filled with large crystals of white calcite or brownish iron-bearing calcite and in places shows drusy openings. In this gangue there are scattered, irregular bodies of galena. The largest body of ore was mined from a tunnel at a barometric elevation of 8,100 feet, on the north side of the hill across which the fissure runs. This body, averaging about 2 feet wide, was 75 feet long on the drift and about 25 feet high on the dip of the vein and pitched northeast in the fissure. At both ends of the ore shoot the barren calcite filling of the fissure continues. About one-half mile south of this body, in the bottom of a gulch, at a barometric elevation of 7,600 feet, tunnels on the fissure open some smaller bodies of ore. There is very little oxidation on this vein. Nearly pure galena outcrops all along its strike, but occasionally small amounts of anglesite and cerusite are seen as thin crusts coating the galena directly at the surface. The sorted galena is said to carry about 2.75 ounces of silver a ton.

Nubagah prospect.—The Nubagah prospect is 4½ miles northwest of Steptoe post office, in the middle part of the east slope of the Egan Range, where the massive limestones are the country rock. A 10-foot shaft is sunk on a tight fissure with north strike and steep east dip. A little galena and cerusite replace the limestone for a short distance on both sides of the fissure. The largest body of ore was found directly at the surface and east of the fissure.

WARD DISTRICT.

LOCATION AND GENERAL FEATURES.

The Ward district (No. 29, fig. 1, p. 18) includes a few square miles of the east front of Egan Range, about 16 miles south of Ely, its shipping and supply point. A good road in Steptoe Valley leads to the camp (barometric elevation, 8,025 feet) at the base of the range, which rises abruptly from the valley floor at least 1,500 feet in about 2 miles. The principal mines are in secs. 14 and 15, though some claims are in secs. 9, 10, and 16, of T. 14 N., R. 63 E. (See fig. 17.) Small springs occur in the vicinity of the mines, and a good stream of water flows in Willow Canyon about 4 miles south of the camp.

GEOLOGY.

This part of the Egan Range seems to consist entirely of a series of thin-bedded, light and dark blue-gray Carboniferous limestones that have been intruded by an intricate system of quartz monzonite dikes. In the surface exposures the igneous rocks are not conspicuous, but in the various mine workings large amounts of them may be seen

SEDIMENTARY ROCKS.

The limestones, which form the greater part of the surface exposures in the Ward district, are in beds 10 inches to 2 feet thick.

They are usually of a light blue-gray color, though some beds are dark gray and some contain a large amount of brown chert. Between the Paymaster and the Mammoth tunnels (Nos. 6 and 8, fig. 17), lime shales are interstratified with the true limestones. Among a few fossils collected by the writer on the trail halfway between the Paymaster and Martin White tunnels (Nos. 6 and 9, fig. 17), G. H. Girty identified the Pennsylvanian forms *Rhombopora* sp., *Productus* sp., and *Spirifer* aff. *S. cameratus.* Spurr[1] reports that a large collection of Permian fossils was obtained from limestones on the southeastern slope of Hamels Peak, whose summit is about 6 miles northwest of Ward. (See Pl. I.)

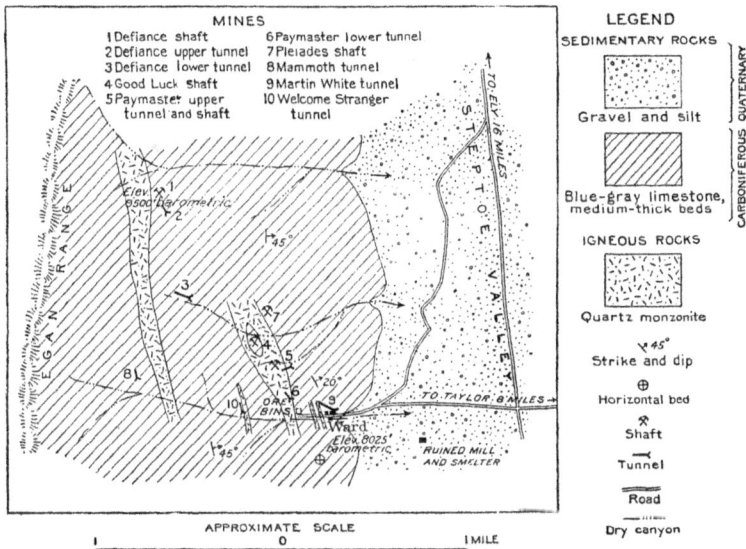

FIGURE 17.—Sketch map of the Ward district, White Pine County, Nev.

In the vicinity of Ward the sedimentary rocks have a north or north-northwest strike and a low east dip. South of the mouth of Ward Canyon the limestones are horizontal. A rather strong fault seems to strike north-northwest across Ward Canyon a mile west of the Ore Bins and east of the upper Defiance workings (Nos. 1 and 2, fig. 17). This fault appears to dip east and is thought to have dropped the beds on the east relatively to those on the west.

IGNEOUS ROCKS.

All of the dike rocks of this district are much altered and most of them have been mineralized to some extent. A specimen of fairly

---

[1] Spurr, J. E., Descriptive geology of Nevada south of the fortieth parallel and adjacent portions of California: U. S. Geol. Survey Bull. 208, p. 52, 1903.

fresh rock from the dump of the lower Paymaster tunnel seems to be a quartz monzonite porphyry, though it may be a granite porphyry. Phenocrysts of a plagioclase feldspar too much altered for definite determination, with some quartz, biotite, and possibly orthoclase are the phenocrystic minerals. The feldspars are altered to masses of calcite and sericite and the biotite is chloritized. The groundmass consists of a microgranular intergrowth of quartz and feldspar, the latter being sericitized. Apatite is an abundant accessory mineral. In most places where seen the highly altered porphyritic rocks retain their texture, though in many places all the minerals, except quartz and apatite, are changed to sericite and calcite. Crystals of pyrite and galena are widely disseminated in the highly altered igneous rocks.

The limestones near the dikes are comparatively free from lime silicate minerals, though in some places they have been silicified and in other places altered to soft iron-stained masses of sericite or calcite. Near the Welcome Stranger tunnel (No. 10, fig. 17) a small mass of dark lime silicate rock lies along the contact between limestone and a porphyry dike, and near the face of the lower Defiance tunnel (No. 3, fig. 17) the limestones are silicified and contain some narrow bands of lime silicate. The most abundant contact-metamorphic mineral in the district is a brownish-gray garnet, though a little light-colored pyroxene, calcite, and quartz are usually present in the zones of lime silicate rocks.

ORE DEPOSITS.

HISTORY AND PRODUCTION.

The ore bodies of the Ward district are said to have been discovered about 1869. The district was organized in 1872, at which time Raymond[1] says that the Martin White Silver Mining Co. of San Francisco owned most of the mines in the district. This company started development in October, 1874, and during the year sank the Paymaster shaft 190 feet. Raymond says that all the mines show considerable quantities of ore, principally sulphurets, assaying from $40 to $300 a ton. Whitehill[2] reports that Ward had a population of 1,000 in 1876, and that a 20-stamp mill and two furnaces of 35 tons' capacity each were in operation. The Austin chloridizing process was used for a time on the rich carbonate ores from the Paymaster and Defiance mines. Whitehill[3] reported that the total

[1] Raymond, R. W., Statistics of mines and mining in the States and Territories west of the Rocky Mountains for 1874, p. 272, 1875.

[2] Whitehill, H. R., Nevada State Mineralogist Sixth Biennial Rept., for 1875-76, pp. 167-168, 1877.

[3] Whitehill, H. R., Nevada State Mineralogist Seventh Biennial Rept., for 1877-78, pp. 160-175, 1879.

production of the district was $550,000 to the year 1878. It is said that the mines were actively worked until about 1882. Plate[1] says that the Martin White Silver Mining Co. did not mine any ore of less than $50 grade, as a consequence of which many tons of fair-grade ore are yet in the mines.

In 1906 practically all of the claims in the Ward district were acquired by the Nevada United Mines Co., which now owns 12,000 acres of patented mining ground and 1,000 acres of ranch land at the mouth of Willow Creek, whose waters they own. According to the estimates of the present owners, $7,000,000 has been taken from these ore bodies. The mines have been idle for several years, but some ore has been shipped from the old dumps. In 1913 a small force of men was reopening the Good Luck shaft and the lower Paymaster tunnel with a view to working the large body of low-grade ore left by the early operators.

OCCURRENCE AND CHARACTER OF THE ORES.

The ore bodies of the Ward district are closely associated with the intrusive quartz monzonite porphyry. They usually occur along the contacts of the igneous rocks as replacements and veins both in the intrusive and limestone. The mineralization took place after the consolidation of the quartz monzonite porphyry, and that rock is everywhere sericitized and calcitized and contains finely disseminated pyrite and galena. The large ore body at the Good Luck appears to be a block of altered limestone included in the porphyry. The Defiance body at the surface is a mass of brown limonite cut by fissures in which rich lead carbonate was found. As seen in the lower tunnel 500 feet below the croppings, it is a mass of silicified limestone about 300 feet wide impregnated with sulphides.

The rich ore mined in the early days was largely argentiferous lead carbonate that carried the silver in part as chloride. The old workings were apparently above the water level. The sulphide zone was cut at a depth of 160 to 180 feet, and the sulphide ores were found to consist largely of sphalerite, pyrite, and galena, with chalcopyrite in some places. As a rule, however, there does not seem to have been a large amount of copper in any of the primary ore bodies.

THE MINES.

Defiance mine.—The Defiance ore body, at an elevation of about 9,500 feet 1½ miles northwest of Ward, is developed by a shaft, now inaccessible, a drift tunnel aggregating 800 feet near the top of the hill (Nos. 1 and 2, fig. 17), and a crosscut tunnel 1,300 feet long, 500 feet below the croppings (No. 3, fig. 17). The croppings are about

---

[1] Plate, H. R., The old camp of Ward, Nev.: Min. and Sci. Press, vol. 94, p. 281, 1907.

200 feet wide and at least 600 feet long along the east side of a quartz monzonite dike that strikes north. They consist of cellular dark-brown hydrous iron oxide that in places shows a little copper stain and in places is cut by northeastward-trending fissures filled with yellow lead carbonate. These streaks of sand carbonate constituted the rich silver ore mined in the early days. It is said that similar ore continued to the bottom of a winze 140 feet below the tunnel level, at which depth a wall that dipped 40° W. cut off the ore. The lower crosscut tunnel runs N. 70° W. for 1,300 feet. Near the end of the tunnel a belt of silicified limestone containing abundant pyrite and some chalcopyrite and galena lies east of a strong fault that strikes N. 40° E. and dips 50° E. Near the fault some narrow bands of gray garnet carry galena and pyrite. Farther east the limestones are silicified and show bands of sulphides as well as the bands containing disseminated minerals. The bands of sulphides, some of which are 20 feet wide, consist almost entirely of pyrite but show a little chalcopyrite. In the bands of silicified limestone, pyrite is more abundant than galena, and both minerals are apt to be segregated to some extent in bands and nodular masses. This whole zone of mineralized rock has been fractured, and little veinlets of calcite carrying secondary galena or a little pyrite permeate it everywhere.

It seems probable that the face of this tunnel is not far from the eastern edge of the quartz monzonite dike exposed on the hill above the tunnel mouth. There is little question that the body of sulphide ore exposed at the end of this crosscut is the extension of the oxidized iron croppings of the Old Defiance workings.

Good Luck shaft.—The Good Luck shaft (No. 4, fig. 17), about half a mile northwest of Ward, is vertical and is said to connect with 3,500 feet of workings. The 280-foot level of this shaft connects with the lower Paymaster tunnel (No. 6, fig. 17). The Good Luck ore body seems to be a replacement of the southern part of a block of limestone included in the large quartz monzonite dike. At the 260-foot level the limestone in an area 60 feet by 150 feet is altered to a soft ocherous mass containing irregular streaks and bunches of hard dark-brown iron oxide and pockets of yellow sand carbonate, which is the high-grade ore. The east, south, and west boundaries of the ore body at this level are porphyry that is bleached and mineralized at the contacts. The north boundary is not definitely marked, the ocher grading through iron-stained rock into unaltered limestone. The ore at this level is said to average less than $10 a ton. On the 160-foot level the ore body is at least 150 feet by 250 feet in area, as shown by a large number of crosscuts and drifts. The east boundary only is definitely marked by the porphyry, which dips 55° W. At this level the bodies of sand carbonate seem to be larger than in the lower level, and some irregular bodies of hard lead carbonate occur.

The ore body at the 135-foot level is about 300 feet long and about 200 feet wide, and is said to average $17 a ton in lead and silver. The ore is redder than in the lower levels and carries some manganese oxide, particularly in the zone between the good ore and barren limestone at the north end of the body.

Paymaster mine.—The Paymaster mine is developed by abandoned shafts and tunnels about a quarter of a mile northwest of Ward and by the Paymaster lower tunnel (No. 6, fig. 17) a few hundred feet west of the camp. This tunnel runs north-northwest for 1,200 feet under the old Paymaster workings to the Good Luck ore body. In the tunnel the northeast-dipping limestones are intruded by a number of small and large dikes of quartz monzonite porphyry. The limestones are somewhat silicified along the contacts but do not appear to be so highly altered as the intrusive rock, which is in many places changed to a mass of sericite and calcite that is thoroughly impregnated with sulphides. The alteration and mineralization of the intrusive rock is particularly strong near a series of fractures that strike N. 20°–40° W. and have a flat northeast dip. The Paymaster ore body is not well shown in this tunnel and the old workings can not be entered. The material on the dump of the upper tunnel (No. 5, fig. 17) indicates that part of the rich ore obtained was an ocherous sand carbonate. Specimens of both limestone and porphyry from this dump carry galena, sphalerite, and pyrite, and one of them carries a small amount of what seems to be argentite and a lead-copper-silver antimony-bearing mineral. A joint in a specimen of the sulphide ore contains a soft greenish-drab carbonate mineral that contains zinc and copper and shows light-purplish and green stains that denote silver and copper and are called silver chloride by the miners. The dumps have been sorted many times and practically all the ore has been shipped.

The dump of the lower tunnel consists of bleached quartz monzonite porphyry and limestone that shows little contact metamorphism. In the ore bins there are several tons of sulphide ore. Black sphalerite, galena, pyrite, and chalcopyrite (named in the order of decreasing abundance) occur in veinlets and irregular replacement veins in both altered igneous rock and limestone but appear to be more abundant in the quartz monzonite. From the poor exposures of the Paymaster ore body in this tunnel it would seem that the original ore minerals will probably be found in the immediate vicinity of the contacts of limestone and porphyry and that the larger bodies will be found in the porphyry, particularly near fissures that strike northwest. Blowpipe and rough chemical tests indicate that the original sulphide minerals do not contain large quantities of silver, and it seems probable that the chief metals found at depth will be zinc and lead.

The Martin White crosscut (No. 9, fig. 17), which starts at the camp, is said to be 3,200 feet long and to undercut both the Paymaster and Good Luck ore bodies. It was caved near the mouth in October, 1913, and could not be entered.

Pleiades shaft.—The Pleiades shaft (No. 7, fig. 17) is about 500 feet north of the Good Luck shaft across a shallow gulch. It is 250 feet deep with three short levels. The small ore bodies are irregularly scattered in a zone of crushed limestone and quartz monzonite porphyry along the footwall of a dike of the intrusive rock which strikes N. 40° W. and dips 50° SW. Oxidation extends to about the 100-foot level, below which the ore is comparatively low grade, consisting of sphalerite and galena, with minor amounts of pyrite and chalcopyrite.

Mammoth tunnel.—The Mammoth ore body (No. 8, fig. 17) is developed by open cuts and a short tunnel. The croppings, which are much like the Defiance croppings, consist of cellular brown limonite. They are 60 feet wide and 300 feet long on the surface a few feet west of a quartz monzonite porphyry dike.

Welcome Stranger tunnel.—The Welcome Stranger tunnel (No. 10, fig. 17) is about half a mile west of Ward on the north side of the canyon. The tunnel is about 500 feet long and trends N. 20° W. It starts 50 feet west of a 50-foot dike of quartz monzonite, which it cuts 200 feet from the mouth. The dike is altered and contains small cubes of pyrite and some galena. The limestone for 30 feet west of the dike is silicified and contains some disseminated pyrite and galena and also narrow bands of lime silicate rock with sulphides. The faulted contacts of the dike in several crosscuts expose gouge and crushed rock that contains sphalerite, pyrite, and galena. A number of east-west veins cutting both the limestone and porphyry carry sulphides and quartz.

AURUM DISTRICT.

LOCATION AND ACCESSIBILITY.

The Aurum district (No. 21, fig. 1, p. 18), as described in this report, includes the north end of the Schell Creek Range, lying in Tps. 20, 21, and 22 N., Rs. 65 and 66 E., and comprises the Muncy Creek, Silver Canyon, Siegel Canyon, and Schellbourne subdistricts. (See Pl. VI.) The Schell Creek Range is long and is relatively narrow in its northern portion. Spring Valley lies east of the mountains and Steptoe west of them. Both of these valleys have typical broad flat bottoms with several playa basins along their axes. The mines and prospects on the east side of the range were visited during the reconnaissance on which this report is based. Aurum post office, at the mouth of Muncy Creek, in Spring Valley, is 42 miles by road southeast of Cherry Creek, its shipping and supply point. Schellbourne, on the west side of the mountains, in Steptoe Valley, is 18 miles southwest of Cherry Creek. A triweekly mail service from Cherry Creek by way of Schellbourne gives easy access to this region. Of late years most of the ore shipped from the east side of the range has been hauled to Currie on the Nevada Northern Railway, about 40 miles north of Cherry Creek. (See Pl. I.) The road over the Schellbourne summit has very steep grades, particularly on the west side of the range. The road northward to Currie has easier grades throughout its length.

TOPOGRAPHY.

In the portion of the Schell Creek Range shown on Plate VI both the east and west slopes are rugged, but the eastern side rises more abruptly from the level floor of Spring Valley than does the western side from Steptoe Valley. Some of the northern part of the area shown is underlain by volcanic rocks, and here smooth gentle slopes prevail. The crest of the range is well above 9,000 feet, and the valleys on both sides are about 5,500 feet above sea level. A peculiar echelon series of indentations on the east front of the range seems to be due to faults which have moved the south blocks toward the east.

GEOLOGY.

The northern part of the Schell Creek Range is composed of a very thick series of quartzites, limestones, and shale, with a prevailing western dip, that have been intruded by dikes and stocks of granite porphyry. The sedimentary rocks have been moved along eastward and northward trending faults, which are believed to be older than the intrusions, though this point was not definitely established. In the northern part of the area both the sedimentary rocks and the intrusives are buried beneath dull brownish and yellowish volcanic rocks which form a considerable group of mountains, extending northeast to the Kinsley and Kern ranges. (See Pl. I.)

SEDIMENTARY ROCKS.

The oldest sedimentary rocks exposed along the east front of the range are dark-weathering dense quartzites that lie in beds 1 to 10 feet in thickness. They are well exposed in the lower parts of Muncy and Silver creeks and along the ridge south of Siegel Canyon. They strike prevailingly north and dip west at medium angles. In Muncy Creek the quartzites are at least 2,000 feet thick, and east of the Siegel mine at least 4,000 feet. In most places they are overlain by dark gray-blue limestones in beds that are 1 to 2 feet thick at the bottom but that a little above become massive, with bedding

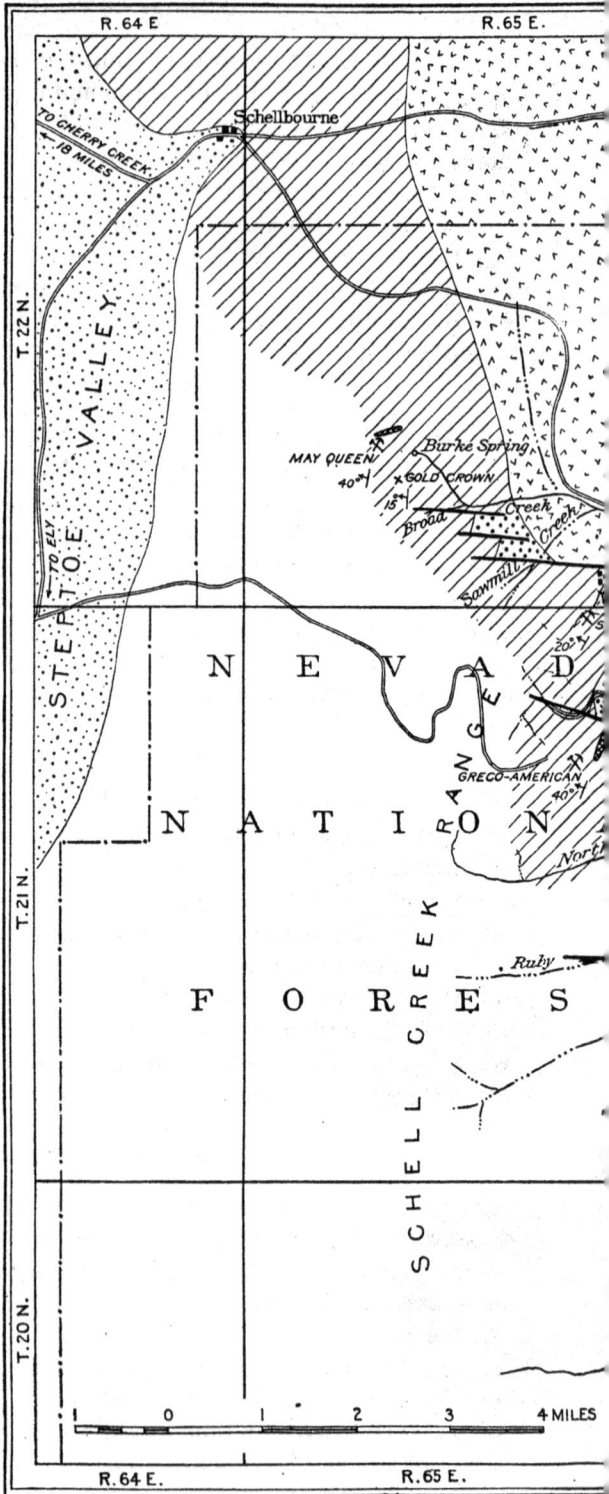

SKETCH MAP OF THE AURUM DISTRIC

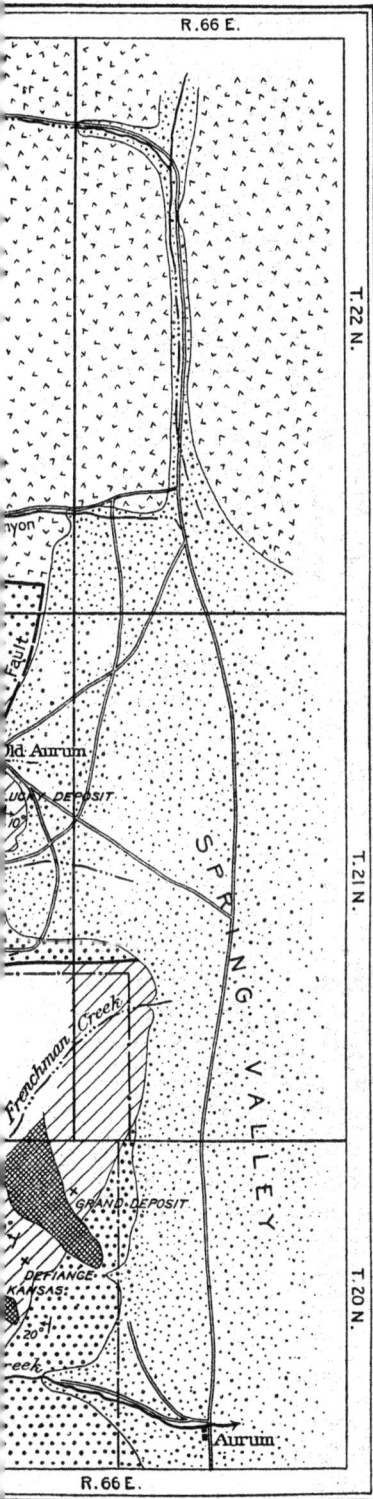

R.66 E.

T.22 N.

T.21 N.

T.20 N.

R.66 E.

nyon

Fault

ld Aurum

UCK- DEPOSIT

SPRING VALLEY

Frenchman Creek

GRAND DEPOSIT

DEFIANCE
KANSAS

20°

reek

Aurum

LEGEND
SEDIMENTARY ROCKS

Gravel and silt

QUATERNARY

Limestone,
dark colored

LOWER PART PROBABLY CAMBRIAN;
UPPER PART MAY BE DEVONIAN

Quartzite, weathering
dark brown

PROBABLY CAMBRIAN

IGNEOUS ROCKS

Andesite and other
extrusive rocks

TERTIARY

Granite porphyry
(intrusive)

LATE CRETACEOUS OR EARLY TERTIARY

Fault

15°
Strike and dip

Mine workings

TE PINE COUNTY, NEV.

planes that are not easily recognized. At the Siegel mine dark shaly quartzites and limestones occupy a narrow zone between the true quartzites and the limestones. In most places, however, the change from siliceous sediments to calcareous deposits is very abrupt, with no intermediate beds of shaly or conglomeratic material. So far as known to the writer limestones form all of the summit of this part of the range and, with some calcareous shales, extend to the base of the foothills on the Steptoe Valley side.

No identifiable fossils were found by the writer in any of the sedimentary rocks in the vicinity of the ore deposits. Gilbert[1] reports that "by the aid of some fossils collected * * * at Ruby-ville and Schellbourne the conclusion is ventured that the base of the series, including the entire quartzite series and the lower portion of the limestone, may be set down as Lower Silurian, while the upper limestones are as recent as Devonian, and perhaps Carboniferous."

Spurr[2] found Cambrian fossils in some dark-blue limy shales on the west side of Schellbourne Pass and correlated the formations on the west side of the mountains with the Cambrian "Prospect Mountain" [Eldorado] limestone, Secret Canyon shale, and Hamburg limestone of the Eureka section. It is the writer's belief that the quartzites and limestones on the east front of the range are the equivalents of the Prospect Mountain quartzite and the Eldorado limestone, of Cambrian age of the Eureka district.

STRUCTURE.

From the east base to the summit of the range the sedimentary rocks have a prevailing strike of N. 10°–30° E. and a dip of 10°–40° W. (average about 30°). So far as noted in this reconnais-sance there is practically no folding of the sediments, which seem to have been tilted by an elevation of the beds along a fault that coin-cided approximately with the east front of the range. This was evidently the interpretation of Gilbert,[3] who says: "This entire series (11,000 feet of quartzites and schists), together with the super-imposed limestones, is tilted in one mass to the west." A portion of the fault is shown south of Old Aurum, separating the limestone of the east foothills from the quartzites forming the mountains imme-diately on the west. A second faulting apparently occurred after the westward tilting of the rocks. These faults trend approximately east, and the movement along them has carried the south blocks toward the east, giving to the east front of the range the echelon ar-rangement already noted. In this reconnaissance only the broadest,

[1] Gilbert, G. K., The geology of portions of Nevada, Utah, California, and Arizona : U. S. Geog. and Geol. Surveys W. 100th Mer. Rept., vol. 3, pp. 30–31, 1875.
[2] Spurr, J. E., Descriptive geology of Nevada south of the fortieth parallel and adjacent portions of California : U. S. Geol. Survey Bull. 208, p. 39, 1903.
[3] Gilbert, G. K., op. cit., p. 31.

most noticeable structural features were mapped, and on Plate VI the positions of only the larger faults are indicated. About a mile southeast of the Gold Crown mine, in Siegel Canyon, there are at least two minor eastward-trending faults, along which there has been some displacement. Some faulting is also suspected northwest of Aurum post office, though no conspicuous breaks are shown by the topography or geology.

It is thought that the intrusion of the granite porphyry occurred after faulting had occurred along both systems, but absolute confirmation of this supposition was not found. One granite dike west of Old Aurum in Silver Canyon appears to continue directly across an eastward-trending fault plane, along which there has been at least 500 feet of displacement.

IGNEOUS ROCKS.

Intrusive rocks.—Intrusive igneous rocks were noted in the vicinity of most of the mines. As a rule the intrusions take the form of dikes that in general strike parallel to the bedding of the sedimentary rocks but do not everywhere conform to their dip. (See Pl. VI.) The largest body of intrusive rock noted is northwest of Aurum post office. Its southern end, a tongue about 1,200 feet wide, penetrates both quartzite and limestone. This rock is a light-gray granite porphyry carrying rounded quartz crystals and orthoclase, biotite, and some plagioclase as thickly studded phenocrysts in a microgranular groundmass that seems to be composed essentially of orthoclase and quartz. No fresh specimens of this rock were found, and the thin sections show a sericitization and calcitization of the feldspars. The border facies of this intrusive is a nearly black glassy rock with minute phenocrysts of quartz, orthoclase, and biotite. The small stock south of the Kansas property is composed of similar minerals, though quartz phenocrysts are not so conspicuous as in the larger mass. The small dikes in Silver Creek are granite porphyries. Some of them contain abundant rounded phenocrysts of quartz, but others are feldspathic, though the microscope shows that they are all quite siliceous.

All of the specimens of intrusive rock collected by the writer are altered, and it is not possible to determine the constituents of the fine-grained groundmass and the plagioclase feldspars. The alteration seems to be of the same type in all places, consisting of a change of the feldspars, particularly plagioclase, to sericite and calcite, and a bleaching of the ordinary dark-brown biotite to a light greenish brown or to a mineral resembling muscovite. Calcite is developed in the groundmass and in the altered phenocrysts, but some calcite has evidently been added to the altered rocks.

The smaller dikes have apparently caused no contact metamorphism of the sedimentary rocks. The quartzites do not appear to have been altered even by the large stock near Muncy Creek; but the limestones in this vicinity, particularly those in the block between the large and small stocks near the Defiance mine 2½ miles northwest of Aurum post office, are altered to masses of yellow-brown garnet, with quartz, calcite, magnetite, and copper and iron sulphides. A little light-colored fibrous pyroxene is developed as radial growths in some of the rock but is not abundant.

Extrusive rocks.—North of Siegel Canyon and extending west along the Schellbourne road to within one-fourth mile of the summit are thick flows of red, brown, purple, and yellow extrusive igneous rocks. Similar rocks cover a considerable area north and east of that shown on Plate VI, extending eastward to the Kern Mountains and north to the Kinsley district. (See Pl. I.) Small isolated areas of volcanic rocks lie north of Schellbourne on the west side of the range and are said to cover a considerable area in the low hills on the east side of Steptoe Valley, south of Schellbourne.

The volcanic rocks in the vicinity of Kinsley are brown, purple, and yellow andesites. Near Siegel Canyon most of the flows are andesitic, carrying phenocrysts of plagioclase, pyroxene, and biotite. In some places the rocks are silicified, so that the groundmass is largely quartz, and some of the plagioclase feldspars have quartz injected parallel to the albite twinning lamellæ. Spurr[1] considered the extrusive rocks at Schellbourne Pass to be pyroxene aleutites. There are, so far as known, no metalliferous deposits in the extrusive rocks, and little attention was paid to their character and distribution during the present reconnaissance.

ORE DEPOSITS.

HISTORY AND PRODUCTION.

According to Raymond[2] the ore bodies on the west side of the Schell Creek Range in the vicinity of Queen Springs and Schellbourne were discovered in 1871, as were also the silver-bearing veins at Piermont on the east side of the range, 15 miles south of Muncy Creek. In these ores silver was the principal valuable constituent, occurring native and as chloride and combined with sulphur and antimony or arsenic. In 1872 Raymond[3] reported that the mines had not proved so good as was expected, as the ores were spotted in the veins, which as a rule were low grade.

---

[1] Spurr, J. E., Descriptive geology of Nevada south of the fortieth parallel and adjacent portions of California : U. S. Geol. Survey Bull. 208, p. 44, 1903.

[2] Raymond, R. W., Statistics of mines and mining in the States and Territories west of the Rocky Mountains for 1871, pp. 200, 203, 1873.

[3] Idem for 1872, p. 168, 1873.

Most of the mines along Silver and Siegel creeks are said to have been located by Simon Davis in the early eighties. The Grand deposit and other properties in the foothills north of Muncy Creek were located by Mr. Noe at about the same time. At one time Aurum and Muncy were flourishing towns, and a 5-stamp mill, located at Aurum, was treating the silver-lead ores obtained from the mines near the head of Silver Creek. In 1913 a little ore was being taken from the Amargosa and Greco-American mines, development was under way at the Lucky Deposit and Queen of the Hills properties, and the ore dump of the Siegel mine was being shipped. So far as known there are no accurate figures of the early production of the mines in the north end of the Schell Creek Mountains. It seems probable, however, that the production has not been large. The Siegel and Lucky Deposit mines have produced more silver than any of the other properties.

Figures of production for 1902–1907, inclusive, collected by the United States Geological Survey, show a production of 74.74 ounces of gold and 245,367 ounces silver, having a total value of $127,163. No production was reported for 1908–1912 inclusive, but in 1913 there was undoubtedly a small production of lead and silver from ores mined on the east side of the range.

OCCURRENCE AND CHARACTER OF THE ORES.

Practically all the ore bodies on the eastern side of the north end of the Schell Creek Range are found in the limestones which overlie the quartzite. At the Signal mine in Silver Canyon the ores occur at the top of the quartzite series only a few feet from the overlying limestones. Most of the ore bodies are replacements of limestone along fractures which trend east. They are somewhat tabular lenses usually parallel to the fissures, but in some places lying along the bedding planes of the limestone. The replacement ore bodies so far developed carry silver and lead, and as most of them are not developed below the zone of oxidation the ore minerals are largely cerusite, though some smithsonite, anglesite, and residual galena are usually present. The Siegel mine has been developed below the oxidized zone, and the original ores consist of pyrite, galena, and arsenopyrite, which occur disseminated in the limestones adjacent to a strong east-west fault. The oxidized ore of this deposit is a soft black material containing principally manganese and iron but carrying silver and some lead.

Northwest of Aurum post office, in the southeastern part of the area shown on Plate VI, small ore bodies of contact-metamorphic origin carry copper, gold, and silver. Pyrite and chalcopyrite with their oxidation products are found directly at the surface of these deposits, but none of the development has penetrated to sufficient depth to reach unaltered sulphide.

At the Signal mine in Silver Canyon pyrrhotite and some chalcopyrite have replaced the matrix of slightly calcareous quartzites and also a small dike of granite porphyry intrusive into both the quartzites and limestone.

So far as could be learned the oxidized lead ores all carry silver, but the ratio of silver to lead varies greatly, some ore carrying very low silver, and other ore, particularly that containing galena, running up to 1 ounce of silver to each unit of lead.

The surface ore of the Gold Crown, Siegel, and Lucky Deposit mines is a soft black manganese oxide which is reported to be the richest ore found in the region. From blowpipe and chemical tests it would seem that what silver this material carries must be largely in the form of chloride.

THE PROPERTIES.

MUNCY CREEK.

Amargosa group.—The Amargosa group of five claims is near the summit of an eastern spur of the Schell Creek Range, about 4 miles northwest of Aurum post office. (See Pl. VI.) In the vicinity beds of blue limestone 27 to 75 feet in thickness are the country rocks. They have a fetid smell, but no fossils were found in them. Bodies of ocherous sandy lead carbonate are opened by two inclines to a depth of 70 feet. A hard reddish-brown iron oxide that is found as a shell about the lead carbonate ores contains a little zinc carbonate as well as cerusite. A carload of sorted lead carbonate ore shipped in 1913 is said to have carried 52 per cent lead and 6 ounces silver a ton. The bodies of ore are irregular and those so far developed are not large. They occur along a more or less open fissure that strikes east and dips 60° N.

Defiance prospect.—The Defiance property about $2\frac{1}{2}$ miles northwest of Aurum, on a low ridge between two small canyons, is developed by an inaccessible shaft and some open cuts along the ore zone. The country rock on this claim is a thin-bedded dark-blue limestone which strikes N. 30° E. and dips 10° NW. The beds are cut by an eastward-trending fissure that dips 45° S., along which the irregular mineralization has occurred. Small masses of the rock have been altered to cellular iron oxide and dark greenish-brown garnet that carry copper carbonates and a little cerusite. The oxidized copper minerals are not abundant in this ore and usually surround grains of pyrite and chalcopyrite. Malachite and a soft red copper pitch ore are the most abundant minerals seen, though a little chrysocolla and azurite are present in places.

Grand Deposit claims.—The Grand Deposit property of five claims is in the lower part of a southeast-draining canyon cut along the east margin of a granite porphyry stock about 3 miles north-north-west of Aurum post office. (See Pl. VI.) The development is through an incline shaft 150 feet deep that dips 60° SW., with short drifts at 60, 100, and 150 feet. These workings are in a zone of intensely crushed iron-stained limestone that contains some pockets of lead and copper carbonates. This zone is about 120 feet wide at the lowest level, at least 50 feet wide at the 100-foot, and 30 feet wide at the 60-foot. On the upper level a 2-foot streak of hard lead carbonate near the east side of the ore zone grades into unaltered limestone. At the surface the west side of the cropping is granite porphyry, but as exposed underground the west limit of the ore is a fracture that strikes N. 50° W. and dips 75° SW.

Kansas claims.—The three Kansas claims are south of the Defiance prospect (see Pl. VI)', near the north contact of a small body of granite porphyry. All of the ores occur in the limestones, which in this vicinity are broken by a number of westward and northward trending fissures. The development work consists of a number of shallow shafts and open cuts on pockets of iron oxide and copper carbonate minerals that are in places associated with brown garnet, epidote, and calcite. A small amount of pyrite and chalcopyrite surrounded by nearly black copper pitch ore is found with contact-metamorphic minerals in several pits. The largest pocket noted was 15 to 20 feet wide and about 200 feet long.

SILVER CANYON.

Greco-American group.—The Silver group of 8 claims, owned by the Greco-American Co., of Ely, Nev., is in the head of Silver Canyon about 2 miles west-southwest of the old deserted town of Aurum. (See Pl. VI.) On the east side of the head of the canyon the main development is a 100-foot incline, which dips 25° N. 35° E., from the bottom of which a narrow irregular drift runs eastward for at least 100 feet along a fissure that dips 70° SE., cutting light-colored crystalline limestones that strike N. 40° E. and dip 50° W. Small, very irregular replacement deposits lie along the fissure, in some places parallel to it and in others forming along bedding planes of the limestone. These ore bodies consist of reddish-brown iron oxide with a large amount of nearly white argentiferous lead carbonate. Some of them contain soft black manganese oxide, though this mineral does not seem to be generally present in the ore. The ore is all hand sorted on the dump and sacked for shipment.

On the west side of the head of Silver Canyon there are some abandoned tunnels on eastward-trending fissures in limestones. The dumps indicate that the ores were very similar to those at the Greco-American mine.

Lucky Deposit mine.—The Lucky Deposit mine is south of the mouth of Silver Creek, about one-half mile from the old town of Aurum (see Pl. VI), on a low hill of dark blue-gray limestone that is separated from the quartzite immediately west by a fault that seems to dip steeply east. The limestone beds near the fault dip 10° W. but east of the mine dip 10° E. They are intersected by numerous open fissures that strike about parallel to the main fault and stand vertical or dip steeply east. Two irregular tunnels which attain a maximum depth of 50 feet constitute the development. They show several lenses of ocherous ore containing hydrous iron oxide and lead carbonate that is said to carry from 40 to 60 ounces silver a ton. Some small bodies of soft brownish-black material in this ore contain manganese and iron oxides that are said to be high-grade silver ore, carrying as much as 200 ounces a ton. The lenses are found both parallel to the bedding and to the fissures, but are most often near the fissures. They vary from 1 to 2 feet in diameter and are connected by narrow fissures that contain a little ore.

Signal mine.—The Signal mine is on the cliffs about 150 feet above and south of Silver Canyon, a mile west of Old Aurum. It is on a strong eastward-trending fault, along which there has been considerable displacement of the beds. (See Pl. VI.) The main ore body is at the top of the quartzite series in beds of somewhat calcareous sandy quartzites that lie below the true limestones. A dike of granite porphyry lies 150 feet east of the ore body, and a large mass of similar rock was seen in the limestone about one-fourth mile south-southwest of the mine. The ore body is developed by two short parallel tunnels 50 feet apart that run west along the main fault and by a 30-foot shaft near their mouths. The cropping of the ore body is fractured quartzite, deeply iron stained, that shows some copper silicates and carbonates. The ore at the bottom of the shaft is not much oxidized. It consists of quartzite and some small dikelets of granite porphyry impregnated with bronze-colored pyrrhotite and minor amounts of chalcopyrite. In thin sections the sulphides are seen to be intergrown and to occur largely as replacements of the calcareous material in the quartzite, though some pyrrhotite apparently replaces the quartz grains. The larger masses of sulphide are associated with a yellowish-brown chloritic mineral and with sericite. This ore is said to carry both gold and copper.

SIEGEL CANYON.

Gold Crown mine.—The Gold Crown mine is in the upper part of Siegel Canyon, about 4 miles southeast of Schellbourne. (See Pl. VI.) It was relocated in 1913 as the Silver Leaf. The thin-bedded dark-blue limestones at the mine strike N. 10°–15° W. and dip 40° W. They are cut by a vertical fissure striking N. 40° W., east of which the ore bodies occurred. None of the old tunnels could be entered, but exposures in some open cuts indicate that the ore occurred as thin lenses parallel to the bedding of the limestone. No ore mineral except soft brownish manganese oxide was found on the dump, and the limestone near the lenses is changed to a black, coarsely crystalline carbonate containing manganese and iron. It is said that the silver occurred with the black or brown manganese oxide and that there was very little lead in the ore.

May Queen group.—The May Queen group of eight claims, formerly called the Burke property, is on the summit of the Schell Creek Range about 3½ miles southeast of Schellbourne. In 1913 this property was under option to H. L. Siegel and F. Falkner, who were doing some development on the main body. The country rock is a blue, somewhat dolomitic limestone in beds 1 to 2 feet thick, which strike N. 10° W. and dip 40° to 50° W. A very much altered feldspathic porphyry dike, seen in indistinct croppings northwest of the ore body, seems to trend northeast and to be 15 to 20 feet wide. The ore body occurs along a fissure that strikes N. 5°–10° W. and dips 80° W. The croppings are 60 feet long and 10 to 15 feet wide and consist of limonite and small amounts of argentiferous cerusite. At a depth of 80 feet residual nodules of galena surrounded by crusts of anglesite and cerusite began to be found in the ore. These nodules are said to assay about 80 per cent lead and 80 ounces silver a ton.

Siegel mine.—The Siegel mine is about a mile south of Siegel Canyon, in the N. ½ sec. 2, T. 21 N., R. 65 E., 6 miles southeast of Schellbourne. (See Pl. VI.) The upper workings are in massive bedded limestones that dip 56° NW. The lower tunnel starts in thin-bedded dark-colored lime shales, which overlie the massive dark-colored quartzite that forms all of the ridge east of the mine. The upper workings consist of a crosscut tunnel which trends S. 10° E. for 100 feet to a fault that strikes S. 61° E. and dips 65° N., along which drifts extend at least 350 feet east and 100 feet west. An incline winze, evidently connecting with the lower tunnel, starts 300 feet east of the crosscut. The lower tunnel starts 350 feet below and about 900 feet north of the upper workings and runs S. 10° W.

The ore in the upper tunnel is black and for the most part rather soft, though some masses of it are hard. It consists essentially of manganese oxides, with some iron oxide and a carbonate that appears to be manganiferous siderite. It is said to carry 40 to 60 ounces silver and 3 per cent lead a ton. Chemical tests indicate that the silver is carried as a chloride.

Sulphide ore found on the dump of the upper tunnel is similar to material on the lower dump, consisting of finely disseminated pyrite, a little of which is arsenical, and galena, in limestone, shale, and quartzite. A little rhodochrosite is present with calcite in this class of ore, which is said to carry about 10 per cent lead and 10 to 20 ounces silver a ton.

WEST SIDE OF THE RANGE.

None of the mines on the west side of the Schell Creek Range were being worked in 1913, and as it was said that most of the old developments could not be entered, the properties at Rubyville, about 4 miles south of Schellbourne, were not visited.

The El Capitan mine at Rubyville is said to be in a 20-foot ledge of white quartz that strikes north and dips west, cutting limestone and shale. The quartz is said to carry argentite with some silver chloride and copper carbonate and to be much like the ore found at the Teacup vein at Cherry Creek.

About three-fourths mile east of Schellbourne an abandoned 200-foot tunnel trends S. 70° E. into faulted limestone. No ore was seen in the tunnel, but it is reported that some small pockets of argentiferous cerusite were found at this property.

DUCK CREEK DISTRICT.

LOCATION AND ACCESSIBILITY.

The Duck Creek mining district (No. 24, fig. 1) covers a low ridge west of the main Schell Creek Range, from which it is separated by Duck Creek, which enters Steptoe Valley about 6 miles north of the smelter town of McGill. (See Pl. I.) Most of the mining properties of the district are on the west side and on the summit, at the northern end of the ridge within 3 miles of McGill, the shipping point. The Success mine is on the divide at the head of Duck Creek, nearly due east of Ely and 18 miles by road from the railroad.

TOPOGRAPHY.

The Schell Creek Range east of McGill splits into two parallel ridges. The higher main mountains lie east of the north-draining valley of Duck Creek and attain an elevation of at least 9,500 feet above sea level. The west ridge, sometimes called the Duck Creek Ridge, is narrow and has a comparatively level summit at an average elevation of 8,200 feet. Its entire west front rises abruptly from Steptoe Valley in steep slopes and cliffs. Duck Creek heads near the Success mine, flows due north for about 16 miles, then turns abruptly west and cuts through Duck Creek Ridge in a narrow, deep canyon about a mile long. All the larger canyons entering Duck Valley head in the Schell Creek Range to the east.

GEOLOGY.

The Duck Creek Ridge is composed of a series of thick-bedded blue-gray, somewhat crystalline limestones, which dip west at medium to low angles. Near the west base of the hills the limestones are overlain by about 300 feet of thin-bedded dark shales, succeeded by thin-bedded dark limestones. Duck Valley is apparently cut from shales which overlie the westward-dipping quartzites and limestones that form the main Schell Creek Range. It is thought that the rocks of the Schell Creek Range are of Cambrian age and that the limestones of the Duck Creek Ridge may be Ordovician and equivalent to the Ordovician Pogonip limestone of Eureka, Nev. No identifiable fossils were discovered during this reconnaissance and the correlation is tentative, but it is concurred in by A. C. Spencer,[1] who paid a hasty visit to this vicinity in connection with his studies of the Ely district. Dense white quartzite that Spurr thought to be the Ordovician Eureka quartzite appears to overlie the supposed Pogonip limestone in the hills about 6 miles south of McGill.[2] It should be understood, however, that there has been considerable faulting in this part of the Schell Creek Range and that it will require detailed geologic work to unravel the stratigraphic and geologic problems.

Igneous rocks appear to be scarce in the northern part of the Duck Creek Ridge. Southeast of the Gallagher ranch, at the mouth of Duck Creek canyon, a small butte composed of a granular igneous rock that is probably a granite rises from the alluvial gravels of Steptoe Valley. A thin flow of rhyolite caps the butte and extends onto the limestones of the hills to the east. On the Ely Gibraltar ground, which lies on the west side of Duck Creek Ridge 3 miles north of McGill, two narrow dikes strike, the one about east and the other northeast. The rocks are too much altered for identification but are thought to be granite porphyries from the few quartz phenocrysts that they contain and from the abundance of orthoclase and quartz in the fine-grained groundmass. The rocks are sericitized and are cut by many veinlets of calcite.

ORE DEPOSITS.

NATURE OF THE ORES.

So far as could be learned the ore bodies of this region have not been known for many years, and probably most of the prospects have been opened within the last 10 years. The Success ore body was discovered in 1907. It is not possible to give the production of the district, as the published figures evidently include the output of mines in the Steptoe or Granite district in the Egan Range.

---

[1] Oral communication.

[2] Spurr, J. E., Descriptive geology of Nevada south of the fortieth parallel and adjacent portions of California: U. S. Geol. Survey Bull. 208, p. 40, 1903.

The ore deposits of this region are of the replacement type, occurring along fissures parallel to the bedding of the inclosing limestone. The mineralization is similar everywhere in the district. The most valuable constituent is lead, which occurs as galena even at the surface, though both cerusite and anglesite are abundant in the ores so far mined. In some prospects copper carbonates occur but are subordinate to the lead ores. The sorted lead ores carry a small quantity of silver, usually less than 5 ounces to the ton.

THE PROPERTIES.

Success mine.—The Success mine is on the divide at the south end of Duck Valley. The workings are in dark blue-gray crystalline limestones that seem to underlie white quartzite and to overlie thin-bedded limestones. About one-half mile east of the mine cliffs of the quartzite are apparently faulted into their present position. At the mine the formations strike N. 35° E. and dip 40° SW., but a short distance to the north similar beds lie horizontal. The mine has not been working since 1910 and access underground could not be gained at the time of visit. The following notes include information supplied by D. C. McDonald, of Ely, who discovered the ore body.

The ore body is developed by an incline that dips east and is 325 feet long, with levels at 110 and 200 feet and some sublevel work. The ore body is 8 to 20 feet wide on the 100-foot level, dips east, and pitches north. The ore seems to occur as bedded replacement of limestone above some thin shaly argillaceous beds. The ore is all of the oxidized type, consisting mostly of sand carbonate, though carrying some massive hard cerusite ore. It is iron stained and contains barite in some specimens seen by the writer on the shaft dump. No sulphides were noted in the ore bins and Mr. McDonald said that none had been found in the workings. The sorted ore is said to average 65 per cent lead, 40 ounces silver, and $10 gold a ton. The ore body has been faulted, and the shaft below the 110-foot level is in limestone to the 200-foot level, which is in shale. The property consists of 22 patented claims, and several ore bodies of similar character are reported to occur at other places on the ground.

Ely Gibraltar claims.—The Ely Gibraltar Mining Co. owns a group of 18 claims covering the west slope and summit of Duck Creek Ridge, about 3 miles north of McGill. The principal development work is a crosscut tunnel, which starts at the edge of Steptoe Valley on the Front claim, trends S. 53° E., and was approximately 900 feet long in October, 1913. Throughout most of the tunnel the massive crystalline limestones, which strike N. 10° E. and dip 50°–70° W., show little disturbance and no mineralization. Near the face of the tunnel a belt of gray crystalline limestone about 50 feet wide contains a little galena associated with masses of white calcite crystals, some individuals of which are 2 inches across. The belt lies above a slip that strikes N. 35° W. and dips 40° NE. A narrow eastward-trending dike 150 feet north of the tunnel is exposed in several cuts.

A few carloads of lead ore have been shipped from an open cut, about one-fourth mile southeast of the mouth of the tunnel. The ore, which is galena surrounded by some anglesite and cerusite, occurs as bedded replacements along a closely spaced series of fissures that strike N. 10°–15° W. and dip 35° E. A few small bodies of copper pitch and copper carbonate ore are associated with the lead minerals directly at the surface, and the presence of zinc is shown by blowpipe tests. This may be the cropping of the body of altered rock exposed at the face of the tunnel, as the ore minerals are similar and large crystals of white calcite are prevalent in both places.

In a second zone of slightly iron stained and coarsely crystalline croppings 150 feet east of the open cut just described, a small amount of zinc carbonate accompanies the lead minerals. Four hundred feet east of the main open cut a 10-foot dike of much-altered intrusive rock strikes N. 51° E. and dips 75° SE. Movement has produced gouge on both walls. A 4 to 6 inch streak of quartz on the footwall side is said to carry about $1.50 gold a ton.

Lead King claims.—The Lead King claims, 14 in number, are on the summit of Duck Creek Ridge, 3 miles east-northeast of Ely, at an elevation of approximately 8,100 feet. Several shallow shafts and prospect pits have been sunk on small bodies of partly oxidized lead ore in the northern part of the ground. The ores occur in lenses parallel to the bedding of limestone which strikes north and dips 45°–50° W. The main work is a tunnel about 500 feet long running southward near the crest of the hill at the south end of the group. A fissure that trends N. 50° E. and dips 60° S. is cut by the tunnel 415 feet from the mouth. Galena and some lead sulphate and carbonate are found in the crushed limestone near the fissure and in irregular bedded replacements near it. The sorted ore is said to average 69 to 72 per cent lead and 3 ounces silver a ton in carload lots.

Mayflower group.—A tunnel on the Mayflower group near the mouth of a canyon directly east of the concentrating plant at McGill trends south and was 1,195 feet long. It follows what appears to be a strike fault in limestones that dip steeply west. Two cross faults which strike N. 70° W. and dip steeply to the south have been followed by short drifts. No ore was noted in the tunnel and none was seen on the dump. An oxidized body of ore on the summit of the hill 1,000 feet above the tunnel level is thought to lie along one of the cross faults cut by this tunnel.

LOCATION AND HISTORICAL DATA.

The Taylor district (No. 28, fig. 1) includes two mining properties which lie close together in the low foothills on the west side of the Schell Creek Range about 16 miles south-southeast of Ely. The mines are about 2 miles east of the stage road from Ely to Osceola (see Pl. I) and are easy of access, both by wagon and automobile. A good-sized town was built in a little canyon south of the mines in the period of greatest activity (1872–1878), but it has been long abandoned. Whitehill[1] reports that the ores carrying galena and copper were discovered in July, 1872, by Messrs. Taylor and Platt, and Raymond[2] says that the Taylor mine was bought in 1875 by the Martin White Co., of Ward, for $14,000. It is said that the total production from the two mines in this district is about 1,000,000 ounces of silver.

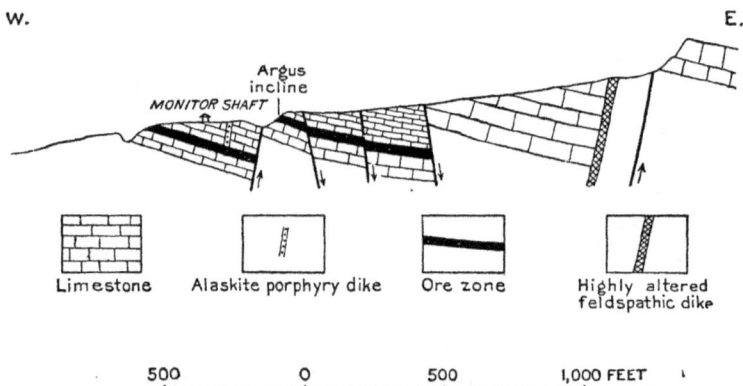

FIGURE 18.—Generalized cross section of the Monitor and Argus claims, Taylor district, White Pine County, Nev.

In 1913 the Argus Mining Co. was prospecting the old Taylor deposit with churn drills. According to this company the ore from the lower body will average $16 a ton, mostly silver but including 60 cents to $1.20 gold. It is the intention of the company to prospect this ground fully and possibly to install a cyanide plant.[3]

GEOLOGY.

In the vicinity of the mines outcropping limestones of undetermined though possibly Ordovician age dip east at low angles. They have suffered a large amount of faulting along a series of northward-trending eastward-dipping planes (see fig. 18), movement along

---

[1] Whitehill, H. R., Nevada State Mineralogist Fifth Biennial Rept., for 1873–74, p. 77, 1875.

[2] Raymond, R. W., Statistics of mines and mining in the States and Territories west of the Rocky Mountains for 1875, p. 194, 1877.

[3] Eng. and Min. Jour., vol. 96, p. 952, 1913.

which seems to have been normal and to have taken place at least in part after the formation of the ores.

A narrow, fine-grained granite porphyry dike, poorly exposed in the eastern part of the Monitor ground, consists of dense white rock carrying small hexagonal quartz phenocrysts about one-sixteenth inch in diameter and a few highly altered phenocrysts that appear to have been orthoclase. The groundmass is a microgranular aggregate of quartz and orthoclase. A second dike about 1,000 feet east of the Argus workings is a soft, iron-stained, highly altered rock, whose original character can not be determined, though it seems to have been a feldspathic rock with considerable quartz.

The Monitor and Argus ore bodies are presumably faulted segments of a once continuous ore body that dips east at low angles parallel to the dip of the sedimentary rocks. (See fig. 18.) The greater part of the Monitor workings were not accessible at the time of visit and it is not known whether any minor faults occur in that body or what the relation of the alaskite porphyry dike is to the ore body. The ore body in the Argus workings lies in a bed of brownish-gray limestone about 70 feet thick below a 15 to 20 foot bed of fine-grained black limestone. The limestone of the ore bed is somewhat brecciated and has been completely silicified along some roughly parallel bands that lie about midway of the upper and lower halves of its width. These siliceous bands, constituting what is said to be the better grade of ore, carrying $70 to $80 per ton, are porous and are lightly stained with copper and lead carbonates. Microscopic examination shows that some dark-gray metallic mineral accompanied the silicification and chemical tests show that this contains iron, copper, and small amounts of arsenic, silver, and lead. On both sides of these bands of high-grade ore silicification and mineralization have penetrated the limestone for varying distances and have formed a low-grade ore said to carry $6 to $10 a ton in silver and gold. This low-grade ore is cut by innumerable calcite veinlets, along which small areas of chrysocolla and cerusite stain are here and there apparent. The ore zone therefore varies from 40 to 70 feet wide with from 4 to 10 feet of barren limestone near the center. In the early days only the two richer streaks appear to have been mined, and the flat stopes on them vary from 3 feet to 6 feet in height. The east dipping ore body has been broken by three normal faults with north strike and steep east dip. The segments east of the faults are dropped 16 feet to 20 feet. Along the fault planes there are some large stopes, one of which, at least 100 by 50 feet in cross section and 60 feet high, must have yielded fairly high grade ore to have been mined in the early eighties, when production costs were higher than at present. Some postmineralization movement has occurred along the faults, but it is thought that the greatest movement occurred prior to the mineralization.

A specimen of sulphide ore found near the west side of the Monitor ground in some open cavelike stopes near the upper part of the ore zone consists of dense black limestone cut by veinlets of white calcite that carry galena, yellowish-brown sphalerite, and a light-gray mineral composed of copper, arsenic, sulphur, and some silver. The three minerals seem to be intergrown, but it is possible that the argentiferous copper arsenide may be a product of enrichment. A small quantity of oxidized ore on the dump of the Monitor shaft carries malachite and cerusite that is shown by chemical tests to be silver-bearing. It is said that the richest ores mined in the early days contained gray copper (presumably enargite), but whether the richness was original in the deposit or was due to enrichment was not demonstrated by the examination of the properties.

KERN (EAGLE) DISTRICT.

LOCATION AND ACCESSIBILITY.

The Kern district (No. 27, fig. 1) includes a large indefinite area in the northeastern part of White Pine County, comprising all of the Kern Mountains and the eastern part of the Antelope Hills. (See Pl. I.) The Kern Mountains are a part of the general uplift of the Snake Range in southeastern White Pine County and of the Toano Range in Elko County. The few prospects in these mountains are widely scattered. Those at the head of Deep Creek are most easily accessible from Wendover, Utah, a town on the Western Pacific Railway, about 115 miles to the north. Others on the west and south sides of the mountains are tributary to Cherry Creek on the Nevada Northern Railway but are also accessible by a practically level road that extends north along Antelope Valley to Dolly Varden siding, about 20 miles north of Currie. Biweekly mail service from Cherry Creek to Tippett station, in Antelope Valley, is maintained by way of Schellbourne and Aurum, a total distance of 66 miles.

GEOLOGY.

During the reconnaissance on which this report is based the southwestern part of the Kern Mountains and the eastern part of the Antelope Hills were visited. The main mass of the Kern Mountains is apparently composed of a fine-grained light-colored muscovite granite that intrudes dark limestones and shales in the southwestern part of the mountains.

On the western and southwestern flanks of Kern Mountains the foothills and lower spurs of the main mountains are formed of dark, thin-bedded crystalline limestones, with some interstratified shales which dip steeply southwest or west. No fossils were found during the reconnaissance, so the age of the rocks is not known, but they are thought to be very old.

Howell[1] noted 200 to 400 feet of quartzites overlying the granite at the head of Deep Creek and underlying shale and 3,000 to 5,000 feet of bluish-gray limestone at Pleasant Valley, both localities being on the east side of the range. He also found Carboniferous fossils at Uiyabi Pass.

Spurr[2] considers this series of quartzite, shale, and limestone to be of Cambrian age, similar to the rocks near Wheeler Peak, in which Cambrian fossils have been found.

IGNEOUS ROCKS.

The granite which seems to form all the central part of the Kern Mountains is a light-colored fine-grained rock which weathers in rounded forms and is clearly distinguishable from the dark sedimentary rocks on the west and southwest flanks.

Concerning the Kern Mountains, Spurr[3] says:

Upon the north side of the Kern Mountains granite is found in contact with the schistose Cambrian quartzites and also with the overlying metamorphic limestones. The central portion of the Kern Mountains is made up of this granite, with the Cambrian rocks on the flanks. A specimen examined microscopically proved to be a biotite-muscovite granite, approaching alaskite. On the borders of the granitic mass are found siliceous granitic dikes which cut the Cambrian quartzite schists. At one locality, which is on the southwest side of Pleasant Valley and near the State line, is found a broad belt of confused alaskite dikes showing a tendency to change into muscovite-biotite granite on the one hand and into larger quartz veins on the other.

Spurr believed that the granite is Archean, notwithstanding the fact that numerous dikes of it appear in the overlying Cambrian sediments.

At the south end of the mountains the granite has an even grain, and few of the minerals which compose it have crystal outlines, though in some places mica plates are fairly well developed. Orthoclase and microcline are about equally abundant and the same is true of quartz and feldspar. Muscovite in small plates is more abundant than biotite. In Water Canyon the granite for about 75 feet north

---

[1] Howell, E. E., U. S. Geog. and Geol. Surveys W. 100th Mer. Rept., vol. 3, pp. 241–242, 1875.

[2] Spurr, J. E., Descriptive geology of Nevada south of the fortieth parallel and adjacent portions of California : U. S. Geol. Survey Bull. 208, pp. 28–29, 1903.

[3] Idem, pp. 26–29.

204 NOTES ON MINING DISTRICTS IN EASTERN NEVADA.

of the contact has a marked schistosity parallel to the contact. The granite breaks into thin slabs along planes in which the muscovite flakes are oriented, and the quartz and feldspar crystals are crushed. Schistosity along the edge of the granite seems to be developed at all places. In Water Canyon on the southwest side of the mountains the normally fine-grained rock grades into either a porphyritic muscovite granite in which orthoclase phenocrysts 2 inches in maximum length are set in a fine-grained groundmass of microcline, orthoclase, quartz, and muscovite, or into a coarse pegmatite of large crystals of all the constituent minerals.

This granite is clearly intrusive into the sedimentary rocks which form the low fringing hills at the southwest side of the mountains. The contact is regular on the whole, but numerous dikes run parallel to the bedding of the sediments near the true contact. Most of these dikes are fine grained, though some are pegmatitic rocks and one shows what appears to be a gradation from granite pegmatite into a quartz vein.

The contact between the igneous rock and the sedimentary rocks on the south and west sides of Kern Mountains is everywhere clearly distinguishable from a distance, for the intrusive rock is light colored and weathers into rounded forms, and the limestones are dark and tend to weather into crags and sharp pinnacles. The contact is marked by saddles on the spurs and by gullies on both sides of the major canyons that cut across the contact. Along the west face of Kern Mountains, near Glenco Spring, the contact seems to strike a few degrees east of north and to dip very steeply west. In Water Canyon, at the southwest corner of the mountains, the contact strikes N. 70° W. and dips 80° S. In the first canyon east of Water Canyon it strikes N. 45° W. and is vertical. The granite is schistose for 100 feet northeast of the contact, and the limestones are crystalline and have a platy fracture for 200 feet south of it. About half a mile south of the contact the thin-bedded dark-gray limestones dip about 45° SW. and are cut by fractures that strike parallel to the contact.

ORE DEPOSITS.

HISTORY.

According to the district records, kept at Tippett post office, the Kern district was organized May 7, 1869, by Messrs. H. T. Fitzhugh, John Eyre, J. Alnathan Smith, E. S. Smith, G. W. Chase, T. A. Stoutenberg, and B. B. Bird. The district as first organized included 10 square miles about Sentinal Peak, which stands near the head of Sentinel Canyon, about 12 miles from Antelope station, on the Overland Stage Route. On May 2, 1870, the boundaries of the district were changed to include all of the Kern Mountains south of old Fort Filmore. At a miners' meeting held July 4, 1872, the district was enlarged to include the eastern part of the Antelope Hills, where the Red Hills mine is located, and the name was changed to Eagle. Most of the claims in the Deep Creek and Pleasant Valley country were located in the early days, and the Well Annie, or Glenco, property was worked in the early eighties. There are no records of production from any of the properties.

CHARACTER OF ORE BODIES.

In the granite core of the Kern Mountains there are quartz veins, some of which, in the southeastern part of the mountains, carry hübnerite, and others, as at Glenco Springs, carry gold and silver associated with enargite, galena, and sphalerite. In a few places small deposits of oxidized lead and copper ores have been found in the sedimentary rocks that surround the intrusive core. None of the deposits have been extensively worked in recent years.

THE PROPERTIES.

Glenco property.—The Glenco group, formerly the Well Annie, consists of 14 claims near Glenco Springs, on the west side of the Kern Mountains, about 7 miles east-southeast of Tippett post office. (See Pl. I.) The property has not been worked for many years and was not visited, as it was said that the 100-foot tunnel and 70-foot winze on the main vein were under water. The contact of granite and limestone runs a little east of north in this part of the Kern Mountains. The vertical vein is said to trend northeast, lying entirely in granite some distance east of the contact, and to be 6 to 8 feet in width. Some ore, said to be from the vein at Tippett, is a white quartz carrying galena, sphalerite, and enargite that is said to carry $8 in gold and varying amounts of silver. Some iron-stained cerusite ore, seen at Tippett, is said to come from undeveloped prospects on the group that lies in the limestone west of the granite.

Red Hills mine.—The Red Hills mine, at the east end of the Antelope Hills, about 15 miles south-southeast of Tippett post office, was worked in 1912 but was idle at the time of visit in 1913. In this vicinity outcropping light-colored reddish-weathering crystalline limestones strike north and dip 30° W. On the west side of the hill a rather thick bed of white calcareous sandstone, which outcrops near the top of the hill, just west of the Red Hills mine, is overlain by light and dark colored, somewhat cherty limestones in beds 1 to 2 feet thick. No fossils were found in the limestones and their age is not known.

The Red Hills mine is developed by an open cut 40 feet long and 20 feet deep at its face; by an inclined winze, apparently about 70 feet deep, 50 feet north of the open cut; and by a 50-foot crosscut tunnel, from whose face a winze is sunk to a reported depth of 200 feet. The winzes could not be entered, for the ladder had been removed when work was stopped. The workings are all in a 20-foot zone of brecciated limestone that strikes N. 10° E. and dips 55°–70° W. and that has been somewhat mineralized near the west wall. At the surface small bodies of iron-stained sandy lead carbonate have been partly removed, but the ore on the dump of the tunnel, presumably taken from the deep winze, is a hard gray lead carbonate.

At the base of the hill about one-fourth mile to the east a 200-foot tunnel is driven S. 80° W. on a fracture that dips 75° S. Near the mouth of the tunnel a small body of brownish sandy material contains some lead carbonate.

Regan tungsten prospects.—The Regan tungsten prospects are on the southeast side of the Kern Mountains south of the head of Pleasant Valley, 3½ miles north-northeast of Mike Spring and about 6 miles from the Utah line. The mines were visited in October, 1914. The veins were discovered in 1910 by C. G. Simms and Casten Olsen and were optioned to O. A. Turner, who was at that time working the tungsten properties near Osceola.[1] At present they are held by Messrs. Simms and Olsen.

The veins, which include nine 2 to 15 feet wide and numerous smaller ones, outcrop over an area approximately a mile wide and 2 miles long on the strike of the veins, which averages N. 20° E. The country rock is a coarse-grained muscovite-biotite granite approaching pegmatite, which is intruded by a series of east-west biotite porphyry (kersantite) dikes and by a few porphyritic biotite granite dikes. The dikes were formed before the veins, for both types are squarely cut by the quartz. The granite is cut by a closely spaced system of joints that strikes approximately N. 20° E. and dips 50° E. The hübnerite-quartz veins are parallel to this jointing, which appears to have determined the position of the veins. The largest vein is developed to a maximum depth of 60 feet by a 200-foot tunnel and by several shafts. In some places it is frozen to the walls, but in most places postmineral movement has crushed the vein filling and produced some gouge on both foot and hanging walls. The granite for a short distance on both sides of the vein is softened by sericitization of the feldspars, and small cubes of pyrite are distributed throughout the altered rock.

The vein filling is largely white quartz in which the metallic minerals are irregularly distributed, usually in small bunches or pockets. The light-brownish hübnerite is the most abundant metallic mineral, but galena, sphalerite, and a small amount of bismuthinite

---

[1] U. S. Geol. Survey Mineral Resources, 1910, pt. 1, p. 739, 1911.

are present here and there in the veins. In vugs in the main vein needle crystals of hübnerite, cubes of purplish fluorite up to one-half inch on a side, and crystals of triplite have been found. Triplite is rare, but the fluorite and needles of hübnerite are more abundant.

Except for the presence of the rarer minerals the vein material resembles some phases of the pegmatitic dikes that grade into quartz veins near Water Canyon, and it is believed that the veins, though of the true fissure type, are the end products of the pegmatitic intrusions.

It is said that the disintegrated granite wash which covers the floor of the valley in which the prospects are situated, contains sufficient hübnerite to wash if enough water were available.

Other prospects.—On the east side of the Kern Mountains in the head of Pleasant Valley some lead-silver and copper-silver deposits are said to occur in blue limestone near the granite contact. These properties were discovered in the early days of the camp but have received little attention, apparently on account of the distance from railways.

In Water Canyon, at the southwest side of the mountains, about 12 miles southeast of Tippett post office, lenses of quartz parallel to the bedding of thin dark-blue limestones near the granite contact have been somewhat developed. A large part of the quartz on the three dumps is barren or contains a little hematite. Some specimens have thin yellow coatings of lead carbonate and one piece of quartz carried a speck of galena.

INDEX.

www.ingramcontent.com/pod-product-compliance
Lightning Source LLC
Chambersburg PA
CBHW070509200326
41519CB00013B/2761